T0271082

The Laser Manufacturing Process

The Laser Manufacturing Process is a comprehensive guide to industrial laser processes, offering insights into their fundamentals, applications across industries, production specifics, and characteristics, including mechanical, metallurgical, and geometrical aspects, as well as potential defects.

The book also investigates how industrial laser processes are developed and the diverse attributes of the resulting objects, emphasizing their significance in industrial settings. Here, "objects" refer to the tangible outcomes of laser manufacturing, encompassing a wide array of products and components created through processes like cutting, welding, and additive manufacturing. These objects exhibit distinct mechanical properties, metallurgical characteristics, and geometrical precision, all of which are crucial considerations in their utility and performance within industrial environments.

This book functions as a concise reference manual catering to the needs of both students and professionals who require knowledge related to laser manufacturing processes, such as laser cutting, laser welding, and laser additive manufacturing processes.

The Laser Manufacturing Process

Fundamentals of Process and Applications

Anooshiravan Farshidianfar,
Seyedeh Fatemeh Nabavi and
Mohammad Hossein Farshidianfar

CRC Press is an imprint of the
Taylor & Francis Group, an informa business

First edition published 2025
by CRC Press
2385 NW Executive Center Drive, Suite 320, Boca Raton FL 33431

and by CRC Press
4 Park Square, Milton Park, Abingdon, Oxon, OX14 4RN

CRC Press is an imprint of Taylor & Francis Group, LLC

ISBN: 978-1-032-76870-0 (hbk)
ISBN: 978-1-032-79466-2 (pbk)
ISBN: 978-1-003-49219-1 (ebk)

DOI: 10.1201/9781003492191

Typeset in Times
by codeMantra

Contents

Preface

The Laser Manufacturing Process holds the promise of transforming the entire manufacturing landscape in the coming decades. No longer confined to prototyping and low-volume production, laser manufacturing processes are now being adopted on a large scale without compromising on versatility. The digitization of manufacturing, personalized on-demand production, distributed manufacturing, and rapid response capabilities during crises have elevated the significance of laser manufacturing processes in the medical and engineering sectors. As industrialized nations strive to regain their leadership in advanced manufacturing through innovation, laser manufacturing processes have become a focal point of research and development efforts. The global economy stands on the brink of the next industrial revolution, with various industries transitioning from traditional production methods to embrace and harness the power of laser manufacturing processes.

However, this promise of the laser manufacturing process comes with its fair share of challenges, particularly in the realm of metal laser manufacturing. Research and development activities are in full swing to address multiple technical hurdles, including speed and productivity, quality assurance, standards, and end-to-end workflow.

One critical obstacle hindering progress in this field is the existing skills gap. Companies aiming to embrace laser manufacturing processes face a scarcity of expertise to guide their entry into the industry. The wider adoption of laser manufacturing processes necessitates bridging the gap in foundational knowledge and understanding within the workforce. Technical experts must possess a comprehensive grasp of laser manufacturing process capabilities to effectively communicate the advantages and limitations to decision-makers while dispelling misconceptions and misinformation. Currently, this knowledge gap significantly impedes sector progress as companies struggle to recruit laser manufacturing process experts who can facilitate effective design development and create meaningful business cases for metal laser manufacturing processes.

This book aims to bridge that gap and help academia and industry overcome these challenges. Strengthening laser manufacturing process skills requires the development of a foundational knowledge base, starting at the undergraduate level. To the best of our knowledge, there is currently no textbook available that connects the fundamentals taught in undergraduate engineering courses with metal laser manufacturing processes. There is a clear need to tailor technical courses related to design, heat transfer, fluid mechanics, solid mechanics, and control to incorporate laser manufacturing process applications. Additionally, business- and management-oriented courses should include laser manufacturing processes to foster considerations of life cycle assessment and business model development among students.

The development of this book was driven by our desire to provide:

1. Foundational material for a core undergraduate course in mechanical and manufacturing engineering.
2. A resource for graduate courses.
3. A concise and accessible reference for industrial engineers, serving as a simple and short dictionary of laser manufacturing processes.

Universities worldwide are revising their curricula to include laser manufacturing process courses. This textbook aims to provide an introductory platform for adoption in such courses, promoting an appreciation for and understanding of the laser manufacturing process among undergraduate and graduate students. Additionally, this book can fill the knowledge gap for engineers working outside academia who wish to comprehend the laser manufacturing process by establishing connections between traditional core physics, engineering concepts, and the laser manufacturing process.

This book offers a step-by-step understanding of the laser manufacturing process, laying a solid foundation for readers to delve deeper into laser manufacturing process research. It focuses on various key laser manufacturing processes, emphasizing basic physics and modeling. This textbook is not a literature survey and is intended for readers with an engineering background. It serves as an introduction to fundamental physical concepts and phenomena in laser manufacturing processes and their applications. Throughout the book, relevant foundational concepts such as laser cutting, welding, hardening, cladding, additive manufacturing, forming, and marking are explored in depth, supported by illustrative case studies.

We would like to express our sincere gratitude to several individuals who contributed to the preparation of this book. Special thanks go to Taha Ghaleb Toos (TGT Co.), who provided valuable assistance with applicable case studies. We also acknowledge the organizations, publishers, authors, and companies that granted permission.

Anooshiravan Farshidianfar
Seyedeh Fatemeh Nabavi
Mohammad Hossein Farshidianfar

Author Biographies

Professor Anooshirvan Farshidianfar is a distinguished figure in mechanical engineering, with a notable academic and industry career. Having earned his PhD from Bradford University in 2010, specializing in Mechanical Engineering, he attained the position of full professor at the Mechanical Engineering Department of Ferdowsi University of Mashhad in 2023. Currently, he focuses his groundbreaking research on additive manufacturing at esteemed universities in Canada and the USA. In 2000, Professor Farshidianfar co-founded Taha Ghaleb Toos (TGT) Co., assuming the role of CEO and steering the company to prominence in the industry. His impactful research, marked by 150 published journal articles, has significantly advanced additive manufacturing knowledge. Recognized as a leading figure in the field, his contributions to academia and industry shape the discipline. Committed to mentoring the next generation, Professor Farshidianfar ensures a bright future for mechanical engineering, solidifying his position as a respected and innovative figure in the field.

Seyedeh Fatemeh Nabavi is an laser material processing researcher. Having obtained her master's degree in mechanical engineering in 2017, she distinguished herself by earning the prestigious Best Thesis Award for her outstanding research work. She has demonstrated her innovation and expertise through the acquisition of national and international patents, with additional applications currently pending. Presently, her research focus revolves around the modeling of characteristics in the additive manufacturing (AM) process.

Dr. Mohammad Hossein Farshidianfar earned his doctorate in Mechanical Engineering—Construction and Production from the University of Waterloo, Canada, in 2017. Following that, he successfully completed a postdoctoral research course in 2018 at the University of Toronto, specializing in Laser Manufacturing. His expertise includes serving as a senior laser material processing engineer in top ten industrial companies, high-impact journals, and leading universities.

Aim and Scope

In *The Laser Manufacturing Process: Fundamentals of Process and Applications*, the captivating nature of light serves as a metaphor to explore the remarkable world of lasers and their significance in industry. This book aims to unravel the mysteries of laser applications by providing a comprehensive understanding of laser manufacturing processes and their diverse applications. By delving into the physics and intricacies of laser processes such as cutting, welding, cladding, forming, hardening, marking, and additive manufacturing, readers will gain a profound appreciation for the invaluable role lasers play in various industries.

This book showcases numerous case studies that highlight the practical applications of lasers in industry, solidifying the reader's understanding of the transformative potential of laser technology. By immersing themselves in the book's contents, researchers, manufacturers, designers, and students working in the field of laser material processing can acquire a deep comprehension of laser manufacturing processes and their practical implications.

This book covers a wide range of topics related to laser manufacturing, including the underlying physics, mechanical, metallurgical, and geometrical characteristics, as well as common defects associated with these processes. It explores the materials involved, theoretical and experimental studies, and models that bridge the microscopic to macroscopic behavior of laser manufacturing processes.

The primary aim of this book is to serve as a concise and accessible reference for professionals seeking clarification on laser manufacturing processes, as well as for scientific scholars and industry professionals interested in gaining a comprehensive understanding of this field. It introduces the fundamental concepts of laser manufacturing processes, their phenomena, and their significance in industry. Additionally, this book offers insights into modeling laser manufacturing processes, discussing various empirical and theoretical modeling methods.

The scope of this book extends to encompass all relevant knowledge related to the design of laser manufacturing processes, presented in a logical order. Whether readers are novices seeking to build a foundational understanding of laser manufacturing processes or seasoned experts in the field, they can rely on this book as a valuable resource. It offers a wealth of theoretical and mathematical deductions, facilitating in-depth analysis, design, and practical application of laser manufacturing processes.

1 Introduction to Laser

1.1 LASER OVERVIEW

Light is a kind of electromagnetic waves that can be utilized and applied in different ways. One of the most promising and impressive ways of serving light is by laser. In other words, high-energy light waves are extracted, focused, and applied in a valuable way by laser technology. From another point of view, it can be said that amazingly, human imaginations and fantasies came true with the invention of the laser.

In a famous science fiction book *The War of the Worlds*, H.G. Wells depicted death by radiation in 1898 [1]. It seems that the origin of the laser came from the science fiction book. The idea of considering the beam as a weapon is also proposed in comic books from the 1950s [2]. Laser technology now provides humans with the ability to imagine using high-energy density beams.

In medical science, lasers are considered to diagnose and treat diseases of the different body organs, including eyes, ears, throat, and nose, as well as various diseases and even cancer cells. It is not an exaggeration to consider laser technology a main fundamental in the revolution of the world of surgery. Thanks to the laser technology, some surgeries that were not possible are now available. Today, laser technology is one of the most powerful tools in different aspects of medicine [3]. Nowadays, because of laser technology, surgeons can perform different types of surgeries in any part of the body with less damage and discomfort to the patient. "Bloodless scalpel" is an example of the use of laser in surgery, which has changed the world of medicine and treatment. One of the most famous surgeries in this field is "LASIK" surgery, which is performed to treat myopia or nearsightedness, hyperopia or farsightedness, and astigmatism. This surgery is performed in 15 minutes without pain or anesthesia. LASIK, which stands for "Laser-Assisted in Situ Keratomileusis," is performed by a femtosecond excimer laser in six general steps. A numbing drop is poured on the surface of the eye (first step). Then a device called a suction ring is placed on the surface of the eye (second stage), and the protective thin layer of the cornea called the flap is separated from it. In this way, the surgeon gets access to the lower surface of the cornea (third stage). In this step, using a laser, very small parts of the cornea surface tissue are removed, and the cornea returns to its original shape (fourth step). Then the flap is returned to its previous state (fifth step). Finally, the cornea of the eye regains its original state.

As discussed, the laser can be applied as a knife during the surgery. The laser-based surgeries have advantages such as a short recovery period, a reduction in pain, no need for hospitalization, and a contact-free laser with the surgical area [4]. Comparing the laser with of the smallest surgical knife, it is revealed that

DOI: 10.1201/9781003492191-1

while the width of smallest knife is in size of a pinhead, the laser can cut the area as narrow as a cell [3]. In laser-based surgery, the infrared beam is completely absorbed by the water in-vivo. Therefore, the tissue placed in the path of the laser beam is vaporized or burned due to the generated heat [3]. Thus, the laser beam, with little bleeding, can create a cut by burning the place [4].

Because narrow beams can pass through the pupil of the eye, the laser-based device is extensively considered in eye surgeries. Before lasers, for any eye surgery, there was a need to cut the pupil [4]. For example, some well-known eye conditions, such as cataracts, glaucoma, and retinal diseases that can cause blindness, can be treated with laser. Argon laser is an example of a laser applied to retinal tear disease [3].

The laser is also applied to the treatment of other organs or parts of the body. For example, the laser can be considered to treat the abnormal growth of some lumps created in the voice larynx [3]. To destroy cancer cells and even treat tow fungus, the laser can be helpful [4]. Nd:YAG lasers are considered to remove cancerous tumors that block the airways of the lungs. This laser is also used in the treatment of bladder cancer [3].

Laser is considered in medicine to treat prostate enlargement. In this method, a laser beam called "photodynamic" removes a part of the prostate that has blocked the urethra. In this method, the surgical area is impregnated with drugs that are sensitive to the laser beam. Then, a scope is inserted by the doctor through the tip of the penis into the duct that transfers urine from the bladder. The prostate is located around the urethra. A laser passed through the scope carries energy that shrinks or destroys the excess tissue that is blocking the flow of urine.

Laser is also applied in dentistry treatments, tools, and parts. Some diseases, such as gum, can be treated by the laser process; for example, it can be applied to treat the tissue of some gum diseases. The laser can quickly treat some oral wounds, such as viral lesions, canker sores, denture wounds, small non-cancerous tumors, etc. [3].

The laser world has also entered mass communication and telecommunication devices. The presence of laser to communicate in space and "hologram" or "holography" are two famous applications of laser in communication devices. Due to the fact that space is vacuum, thanks to laser, it is possible to communicate between satellites that are several thousand kilometers away from each other's. The laser can also transmit information from satellites to airplanes. This laser technology is called "LLCD," or Lunar Laser Communication Demonstration. The LLCD system is currently at the disposal of the American Space Agency (NASA) and is the first dedicated communication system that, unlike radio waves, provides the capability of a huge amount of information for long-distance communication. Thanks to this technology, astronauts will be able to take higher-quality images and even 3D video images from the infinite space and transmit them to the earth through the connection between the LLCD and its receiver terminal, or Lunar Lasercom Grounded Terminal (LLGT). It is worth noting that it has been stated that by 2030, communication between Mars astronauts will be carried out with this technology; the amount of information sent is six times that of radio waves (with a speed of 600 Mbps).

A hologram[1] is a three-dimensional image created by recording light and converting it into an image. Perhaps the most suitable light for creating a 3D hologram is laser light. Because its beam is coherent and continuous. Helium-neon laser with a 633 nm light spectrum are often applied in holographic systems. In a hologram, in the first step, the laser beam is irradiated on the two-dimensional image. Then, with the refraction of light (which is usually divided into two parts), two images of two dimensions will be created. Finally, a three-dimensional image is formed from the intersection and integration of two-dimensional images.

In scientific research and studies, the presence of the laser can be seen with "spectroscopy," "laser cooling," "LiDAR," or "remote sensing." Spectroscopy, to which mankind owes its knowledge in the field of molecules and atoms, is the study of the interaction of light and matter. In this method, the material is exposed to light. The necessary information will be extracted from the examination of the frequency spectrum reflected from the material [5]. To study more about the details of laser spectroscopy and its different methods, you can refer to the research [6,7].

The laser cooling process refers to the methods by which the atoms and molecules of the object are cooled to near absolute zero by the laser. In this method, atoms and molecules are stimulated with a laser, and as a result, their momentum changes. Because the temperature of particles is a function of changing their speed or momentum. Therefore, by adjusting their momentum, the temperature of molecules and atoms can be cooled to absolute zero. To understand more details, you can refer to studies [8,9]

LiDAR, which stands for "Light Detection And Ranging", is one of the remote sensing methods in which lasers are used. In this method, light pulses are irradiated on the object by a laser. By measuring the beam reflections with the sensor, the distance and three-dimensional geometry of the object can be obtained. In some cases, LiDAR is called a 3D laser scanner. This method is used in sciences such as geography [10], geology [11], seismology [12], and cosmology [13]. This technology is also used in the control and routing of unmanned vehicles [14].

In addition to the world of research and development, the presence of lasers is also observed in commercial products. Printers, barcode scanners, and pointers are three well-known uses of lasers in commercial products [2]. With all these examples, the extent of laser application in any field is not as impressive and significant as the industry. There are various reasons involved in the occurrence of this issue, all of which will be mentioned and listed in detail in this book. In this regard, it might be better to first familiarize yourself with the history of the laser and its governing physical laws. Therefore, in the next section, the general history of lasers will be presented.

1.2 A BRIEF HISTORY OF LASER

Some consider the beginning of the laser theory to be the presentation of Maxwell's equation by James Clerk Maxwell in 1865 [15]. He provided equations to describe the nature of light waves and believed that electromagnetic waves travel exactly at the speed of light. It should be noted that Maxwell believed that the behavior of light was wave-shaped. Later, scientists observed that light behaves sometimes as a particle and sometimes as a wave [16].

Some materials can absorb radiant energy and emit it as light. This condition occurs when the stable electron of an atom absorbs some radiation energy and moves to the excited level. The excited electron, which tends to reach a stable state, releases some of its own energy during its return, which is emitted as light. In his studies between 1905 and 1916, Einstein called the basic unit of light emitted during emission "Photon." His purpose for this naming was that indivisible packages of light, "Quantum," behave like moving particles. He believed that a photon could be completely absorbed by an electron. Also, in a further perspective, he emphasized that a photon emitted from an atom can create another similarly excited atom so that the second photon is emitted. He called this form of emission "Stimulated emission" [17]. With this explanation, some mathematical studies call Einstein the initiator of the laser theory [18].

According to the studies of Maxwell and Einstein, the practical reality of the laser remained silent until 1951. This year, Charles Hard Townes and his colleague Gordon Gould, who were looking for a way to increase the range of microwave frequencies, mentioned Einstein's theory. Their goal was to increase the frequency of microwaves in order to use them in the communications of the US Navy. He considered a chamber with a high gas temperature; some of its atoms were in an excited state, and some were in a stable state. They made the molecules inside the chamber into excited molecules with high energy that can achieve optical resonance. In this situation, the excited molecules could be excited by a signal with a certain frequency so that they emit microwaves. If the walls of the desired pocket or optical resonator can reflect the beam well, the excited atoms can stimulate other atoms, and as a result, more chains of microwaves are emitted. In 1953, this proposal was made using ammonia gas molecules as the active medium inside the resonator, and it was called "Maser." This term stands for "Microwave Amplification by Stimulated Emission of Radiation" [19].

After the discovery of the maser, in 1958 Townes, in collaboration with Arthur Leonard Schawlow, built a maser that produced microwaves with shorter wavelengths, which eventually led to the emission of visible light instead of microwaves. They called this device "Laser," which means " Light Amplification by Stimulated Emission of Radiation" [15,20].

The above-mentioned was a summary of outstanding research in the field of laser history. The studies on laser was continued, and various researchers were provided, which led to the current position of laser in modern technology. Masses of researchers and scientists made and applied other materials as active laser medium. A brief history of laser was provided and is presented in Table 1.1.

TABLE 1.1
The Historical Pass of the Emergence of Lasers

Year	Presenter	Achievement	References
1865	Maxwell	Maxwell's equation	[14]
1888	Heinrich Rudolf Hertz	Generation and detection of electromagnetic waves	[14]
1900	Max Planck	Quantization of radiation in a cavity	[14]
1905	Einstein	Quantization of radiation	[14]
1905	Niels Bohr	Quantization of the energy states of an atom	[14]
1916	Einstein	Theory of light emission and presenting the concept of stimulated emission	[4]
1917	Einstein	Spontaneous and stimulated emission	[4,14]
1923	Kramers	Influence of stimulated emission on the refractive index of atomic gases	[14]
1928	Ladenburg	Observation of an influence of stimulated emission on the refractive index of a gas of neon atoms excited	[21]
1940	Fabrikant	Proposed population inversion	[4]
1947	Lamb and Retherford	Induced emission suspected in hydrogen spectra	[17]
1951	Townes	Idea of a maser	[4,14]
1951	Weber, Prokhorov, Basov	Inventor of the maser	[4]
1954	Dicke	Optical bomb patent based on pulsed population	[18]
1956	Bloembergen	Proposal of the three-level maser	[19]
1957	Gould	First document defining a laser	
1958	Schawlow and Townes	Proposal of infrared and optical masers (lasers) including the formulation of the threshold condition of laser oscillation	[15]
1959	Basov	Proposal of the semiconductor laser	[14]
1960	Schawlow, Townes	Laser Patent No. 2,929,922	[20]
1960	Maiman	Ruby laser	[22]
1960	Sorokin and Stevenson	Uranium laser	[23]
1960	Javan, Bennett, and Herriott	Helium-neon laser	[24]
1962	Hall	Semiconductor laser	[25]
1963	Kroemer	Proposal of the heterostructure laser	[14]
1964	Geusic et al.	Inventor of the first working Nd:YAG laser	[26]
1964	Patel	Carbon dioxide laser	[27]
1964	Bridges	Argon ion laser	[28]
1965	Pimentel and Kasper	First chemical laser at	[29]
1966	Silvast et al.	Metal vapor laser	[30]
1977	Madey et al.	Free-electron laser	[14]
1979	Walling	Alexandrite laser, the first tunable solid state laser	[14]
1982	Moulton	Titanium-sapphire laser	[14]
1991	Hasse et al.	Green diode laser	[14]
1994	Capasso et al.	Quantum cascade laser	[14]
1997	Nakamura	Blue diode laser	[14]

It should be noted that studies and research related to laser do not end in 1997. After this year, laser research expanded into various fields. For example, the history of laser in skin has been investigated in a study [31] and its history in dentistry in a study [32]. Research related to laser scanners has also been reviewed in recent years [33]. An overview of pulsed laser Nd:YAG technology used in drilling was studied in another study [34]. Advances related to fiber lasers have been reviewed and presented in another study [35]. According to the large volume of studies conducted in the field of laser, in order to understand their importance, their characteristics will be investigated further.

1.3 LASER CHARACTERISTICS

A laser is a device that emits light in the form of very narrow parallel beams that have a specific wavelength. In the first step, the question may arise: what is the difference between the light emitted by a laser and a lamp? Studies show that lasers do not produce much power in most cases. Comparing the laser with a normal 1,200 W hair dryer, it can be seen that the hair dryer is more than 90% more powerful than today's lasers. Experiments reveal that, in most cases, more than 90% of the energy of the laser device is wasted. Of course, diode laser with 70% efficiency are an exception [36]. The fiber laser efficiency is about 100% [37]. According to the mentioned example, the question arises: basically, why laser? What makes laser so attractive that they have grown so fast in the industry? What makes laser technology so special? In the following, by stating the contents presented in the following sections, the reader can decide on the answers to these questions.

1.3.1 Monochromatic

The first difference between laser technology and other light rays is related to the wavelength. While the light emitted from the lamp contains several wavelengths (white light), the light caused by the laser has only one specific wavelength. This laser property is called "Monochromatic" or "single wavelength" in physics [3,36]. Although it is said that the laser is single-wavelength or single-frequency, it cannot actually be assigned an exact frequency. For this reason, some call this property "small frequency band" [38]. To understand this feature, look at Figure 1.1.

In part a of Figure 1.1, the white light emitted from the lamp is placed in front of the prism, and the output of the prism includes light waves with different wavelengths. It should be noted that different colors of light waves represent different wavelengths. On the other hand, according to part b of Figure 1.1, it can be seen that the laser light output from the prism is single-color and single-frequency [36]. Therefore, the difference between white light and laser light is understood from the wavelength point of view.

1.3.2 Directionality

Another difference between laser light and lamps is in the direction of their radiation. The lamp emits light in all directions, while the laser emits light

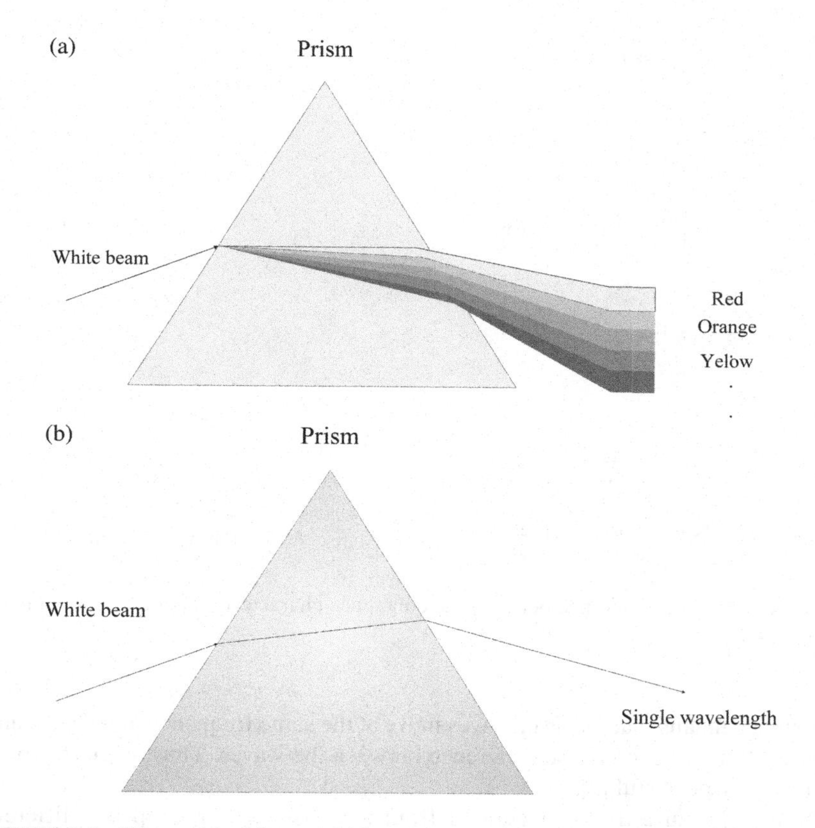

FIGURE 1.1 Waves: (a) white and (b) laser single wave in the prism.

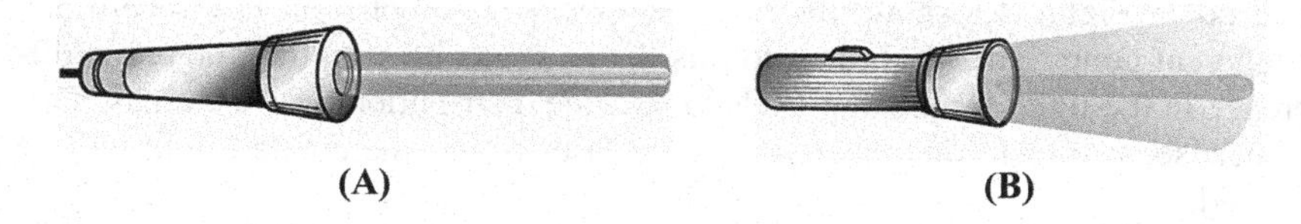

FIGURE 1.2 Output light: (a) laser and (b) flashlight.

in a specific direction. Therefore, the laser light is "directional" and parallel [14]. For a better understanding, in parts A and B of Figure 1.2, the light output from the laser and flashlight can be seen, respectively. Like the previous feature, this feature is also not absolute, and for this reason, according to some, it can also be called "low beam divergence" [38].

1.3.3 Coherence

Another characteristic of the light emitted from the laser is its "Coherence," due to which the energy loss of the laser beam is less than that of the lamp beam [38].

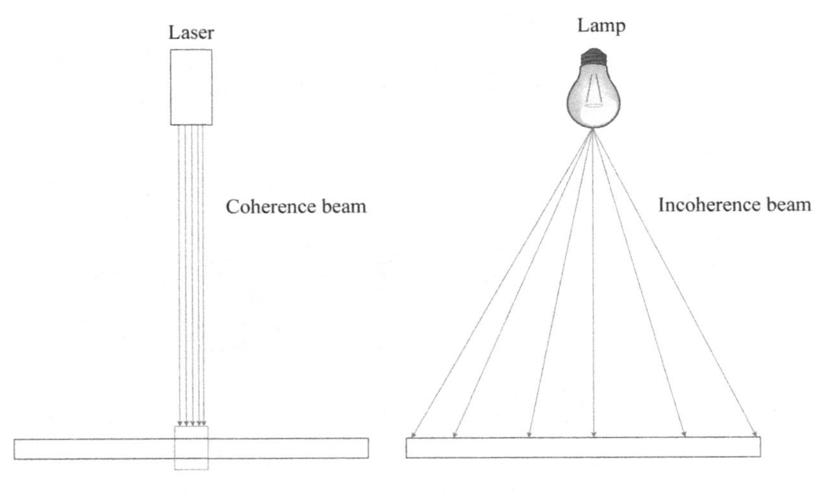

FIGURE 1.3 The difference between the coherence characteristics of a laser beam and a lamp light.

Coherence means that the laser waves move at the same frequency and phase, and as a result, there is no phase difference between the waves. Thus, its narrowness will be very powerful [3].

This is because the light emitted from the flashlight or lamp has different phases. Therefore, it can be seen that, thanks to the coherence feature of the laser beam, all the rays emitted from it are irradiated without phase difference from each other. Figure 1.3 shows the difference between a coherent laser beam and incoherent lamp light. According to this figure, it can be seen that the phase difference in the light of the lamp leads to its wide distribution over the areas. This is because, thanks to the coherence of the laser beam, the created light will be focused.

1.3.4 High Energy Intensity

Another feature of the laser, which can be considered a result of the previous three features, is the high intensity of energy at the focus point. Some also introduce this feature with the name "brightness" [39]. According to the previous three characteristics that can be seen in Figure 1.4, it can be expected that laser waves that are moving with the same wavelength, in a specific direction, and without phase difference relative to each other have a high energy intensity value at their focus point. to have In a study, the amount of energy intensity at this point was reported to be about 1,015 W/cm^2 [38]. According to these unique features of laser, the main components of laser and how to produce them will be presented below.

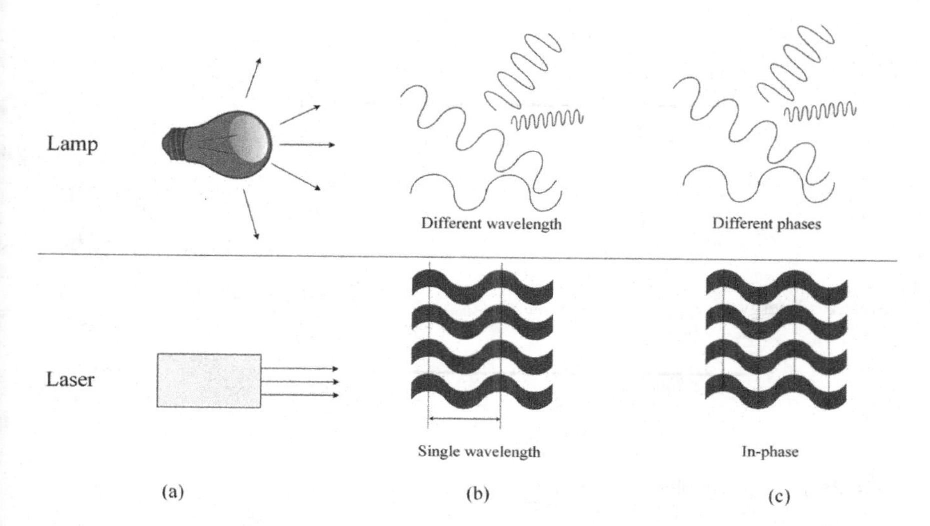

FIGURE 1.4 Laser light characteristics: (a) directionality, (b) monochromatic, and (c) coherence.

1.4 LASER AND ITS DIFFERENT PARTS

Laser consists of three main parts, which are (1) the active medium, (2) an external energy source or blower, and (3) a housing or optical resonator [4,40]. For better understanding, the schematic view of laser components is shown in Figure 1.5.

According to Figure 1.5, the laser is created in several steps. According to number (1) of Figure 1.5, the energy source or pump emits energy in the form of

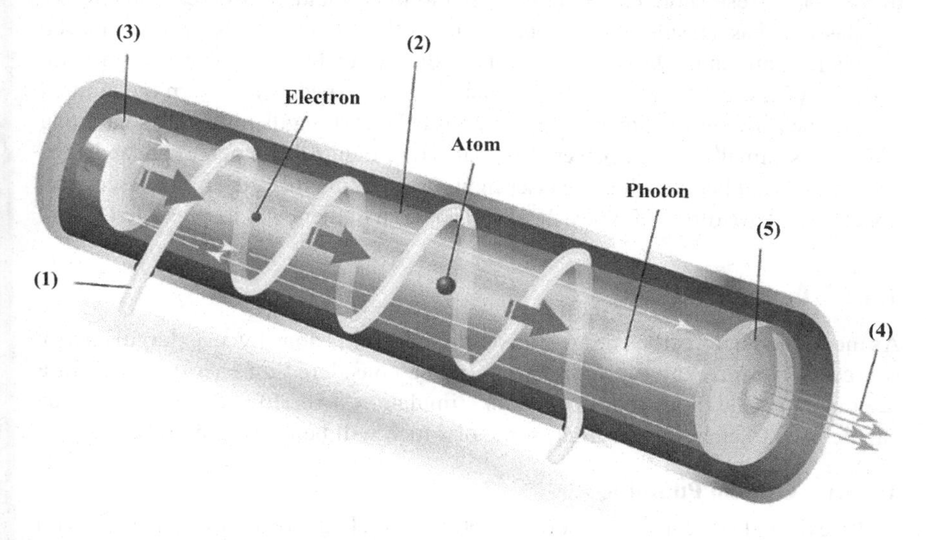

FIGURE 1.5 The schematic of the laser.

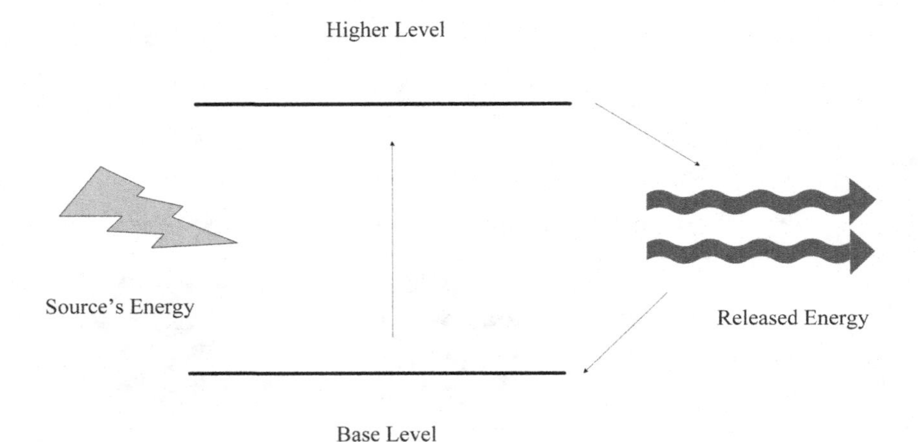

FIGURE 1.6 The molecular behavior of a laser.

light flashes into the active environment. The active medium, shown by number (2) in Figure 1.5, absorbs this light. Then, some electrons in the active environment are excited and go to a higher energy level. These electrons, which tend to return, release their own energy, a schematic view of which can be seen in Figure 1.6. The energy released, or the repeated flashes of light from the energy source, continues. When the number of excited atoms in the active medium exceeds the number of stable atoms, "Population Inversion" occurs. Population inversion is essential for laser [3].

Note that the released energy will be in the form of a light wave that is trapped in the optical resonator. This light will lead to the excitation of other atoms in the resonator, and as a result, light is released from other atoms, and as a result, induced emission occurs in the laser. The mirrors of the two ends of the resonator, parts (3) and (5) of Figure 1.5, reflector and output coupler, respectively, return the emitted light to the active environment. This action is called "strengthening." By continuing this process, finally, a high-power light, part (4) of Figure 1.5, emerges and exits the hole of the chamber, which is the laser light. Considering that each of the laser components can have different types, the details of each of them will be presented below.

1.4.1 Pumping Source

As mentioned before, the energy transfer to the laser is done by an external energy source or blower. Any type of energy can be considered an energy source in a laser and provide the energy needed for stimulation. Therefore, different methods can be used to energize the laser, some of which will be introduced below.

1.4.1.1 Optical Pumping

In the external optical energy source method, which is mostly used in lasers with solid and liquid active media, light energy is used to excite atoms. In other words,

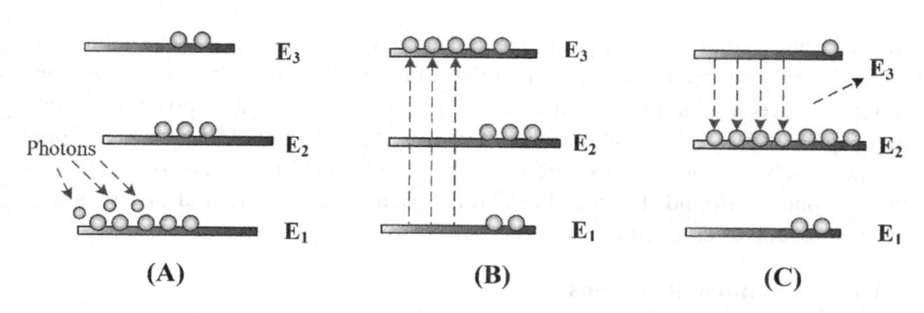

FIGURE 1.7 The method of population inversion in an optically driven laser: (a) the phase of receiving energy from the energy source, (b) the phase of atoms going to the excited energy level, and (c) returning from the excited energy level to a new level.

when a photon or light energy enters the active medium, the atom will receive the photon energy and go from the ground state to the excited state. It should be noted that the required light energy, or photon, is provided by an optical source [41]. For a better understanding, you can pay attention to the schematic in Figure 1.7.

According to part a of Figure 1.7, the photons collide with stable electrons at the E1 level of the atom. This light energy is absorbed by three electrons, and these electrons go to a higher energy level, E3 (part b of Figure 1.7). Excited atoms do not stay long at the E3 level. After a short time, they release some energy and go to the nearest energy level, E2 (part c of Figure 1.7). The electrons at the E2 level have a longer life than at the E3 level. This is because they have more energy than the E1 level. Therefore, the inversion of the population occurs with this method in the laser.

The flash lamp is an example of an optical source used in the laser beam. A flash lamp is a lamp that can produce high-power white light in a short period of time using electric current. The flash lamp is made of a glass tube containing a gas, for example, Xenon, and electrodes are placed inside it [42]. Be careful that a capacitor is needed in lasers with a flash lamp blower. Because of the capacitors, electric discharge takes place in the flash. As a result, a lot of energy is produced from them. This energy is the same photon required for the excitation of the active medium [43]. A flash lamp is applied to ruby laser and most pulsed lasers [41].

The arc lamp is another example of an optical blower that is often applied in the blowers of solid and continuous lasers. In the arc lamp, light will be produced with the help of ionization between two electrodes and the production of an electric arc. The light from this lamp contains a large amount of ultraviolet radiation. Krypton is usually applied as the gas in arc lamps. Because the output power of this gas is much higher than others, it is worth mentioning that while the Xenon flash lamp generates energy in pulse form, the energy generated from the arc lamp is continuous [42].

Another example of optical energization is using another laser as a blower. At first glance, it may seem a little strange! Basically, if you need a laser as a blower for another laser, why not use the primary laser? The reason for this is the increasing efficiency of this type of optical blower. Considering that the light output from the laser has a small wavelength range and is almost single-wavelength, its use as a laser energy source is far better than that of the flash lamp, which emits a wide range of wavelengths. For example, diode lasers are applied as blowers for solid and liquid lasers. The wide application of this type of optical blower is in fiber lasers, which are widely considered in today's industry. In fiber lasers, the diode laser is placed around the fiber laser in the form of a coating and provides the energy needed to create the laser process.

1.4.1.2 Chemical Reactions

If a molecule or atom is produced during chemical reactions and remains in its excited state after production, it can be used as a laser blower. In other words, in this method, by using some chemical reactions, the atoms of the active environment can be excited. Consider the active medium of the laser liquid. By adding some chemicals, due to some chemical reactions, the atoms go from a stable state to an excited state. When these atoms return to their stable state, laser light is created. In this method, the particles of the active medium are raised to a higher level by chemical exothermic interactions. These types of blowers are used in manual and excimer lasers [4].

1.4.1.3 Electrical Discharge

The use of electrical discharge as a blower is common in gas lasers. For example, in the helium-neon laser, the electrons caused by the electric discharge collide with the helium atoms and excite them. Excited helium atoms collide with neon atoms and transfer energy to them. As a result, population inversion occurs in neon atoms. This method is usually used in diode lasers and semiconductor crystals (e.g., germanium). The mentioned cases were common and widely used methods in the field of laser blowers. In addition to the mentioned methods, other methods, such as microwaves [44], can also be used to excite the active environment. For further study, you can refer to the mentioned sources.

1.4.2 Active Media

A set of atoms, molecules, ions, and, in some cases, semiconductor crystals together form a material called "active medium" for the production of the laser process [45]. It should be noted that any material, including solid, liquid, gas, or plasma, can be used as an active medium. Because every medium has the ability to emit energy if stimulated properly [4], it is enough to understand the vastness of the laser world that even dessert can be used as a laser environment [46].

The output power of the laser depends, to a significant extent, on its active environment. Therefore, by controlling the conditions of the active environment, the power of the laser can be controlled. For example, in a study, the effect of material flow direction on the amount of laser power produced was investigated. In this study, under the conditions of fast and axial flow of carbon dioxide, laser power generation of 0.7 kW/m was reported. In this situation, the laser power in the conditions of this active environment was observed in calm or static conditions, with a slow or static current of 0.05 kW/m. Therefore, not only the material of the environment but also its direction and speed affect the laser power [4]. In the following, the details of each type of active environment for lasers will be provided.

1.4.2.1 Gas Laser

In gas lasers, the gas flows inside the resonator. When the gas flows, it passes through two electrodes, one positively charged and the other negatively charged. The electrons flowing between the electrodes pump the electrons of the flowing gas atoms to high energy levels [3].

Note that unlike solid lasers, gas lasers can flow in the resonator. As the gas flows, the excited atoms move to a lower level as they move away from the electrodes. Electrons from excited gas atoms emit photons by going to a lower energy level. These photons are reflected back and forth between the mirrors and thus amplified. When the energy level reaches the required level for the laser, its beam is removed from the cavity [3]. Different gases can be used as active media in lasers. Examples of these lasers are carbon dioxide, carbon monoxide, helium-neon, helium-cadmium, argon, grypton, molecular nitrogen, exmir, copper vapor, and gold vapor [47].

1.4.2.2 Solid Laser

In order for a solid crystal to be used as an active laser material, it must have certain properties. The crystal must be transparent so that light can enter it to excite the environment. Also, this transparency should be such that the laser beam can pass through it. The active medium of the laser can be different solids. An example of solid lasers are agate laser, neodymium yag, and fiber [3,14].

Solid lasers are not very efficient. Energy transitions that occur in a solid-state laser generate heat. In order for solid lasers to have time to cool down, unlike gas lasers that produce a continuous light beam, they work in pulses. On the other hand, solid-state lasers can produce extremely powerful pulses of laser light. For example, the largest neodymium lasers can generate a power of 25 trillion watts in one pulse [3].

1.4.2.3 Liquid Laser

In some lasers, liquid is used as an active medium. Considering that liquids are denser than gases, liquid lasers have the advantage that they can be cooled more easily by flowing. Color lasers are an example of these lasers. Colored lasers such as fluorescence can emit a laser beam with different wavelengths, so these lasers are tunable. For example, the wavelength of color lasers can range from 350 to 1,800 nm. With this advantage, the consumer can adjust the laser device to the desired wavelength according to their needs. Also, the beam from liquid lasers can be emitted discretely and in femtoseconds. This point is another advantage of liquid lasers [3].

It should be noted that color lasers are dissolved in a solvent such as alcohol [3], ethylene glycol [3], or water [4]. In these lasers, the output source is a flash lamp or another laser. For example, Rhodamine 6G laser can be used as a color active medium, whose energy source can be an argon, or krypton ion laser. The beam produced by this laser can include a wavelength ranging from 570 to 655 nm [3]. Coumarin 2 and Coumarin 30 are two other examples of dyes used in liquid laser [4].

1.4.3 Optical Resonator

An optical resonator, or optical cavity, is a chamber in which a laser light beam is produced from the active medium [4]. Therefore, the optical resonator is the main part of the laser. In order to intensify the light wave, lenses with different arrangements are installed on both sides of the chamber. Thus, the light wave produced in this environment between the lenses acts in a reciprocating manner and is somehow intensified [4].

According to the focal radius of the lenses and their distance from each other, there are different types of optical resonators, which will be described below. If both lenses are of flat type and their focal radius is infinite, the resonator will be of parallel-plane type. In part a of Figure 1.8, the schematic view of this resonator can be seen. If the focal radius of two lenses is equal to half the distance between them, it is called a spherical or concentric resonator (part b of Figure 1.8). A confocal resonator, part c of Figure 1.8, is a resonator in which the focal radius of the lenses is equal to the distance between the lenses. In part d of Figure 1.8, the hemispherical resonator can be seen. In this resonator, one of the lenses is flat, and the focal radius of the other lens is equal to the distance between the lenses.

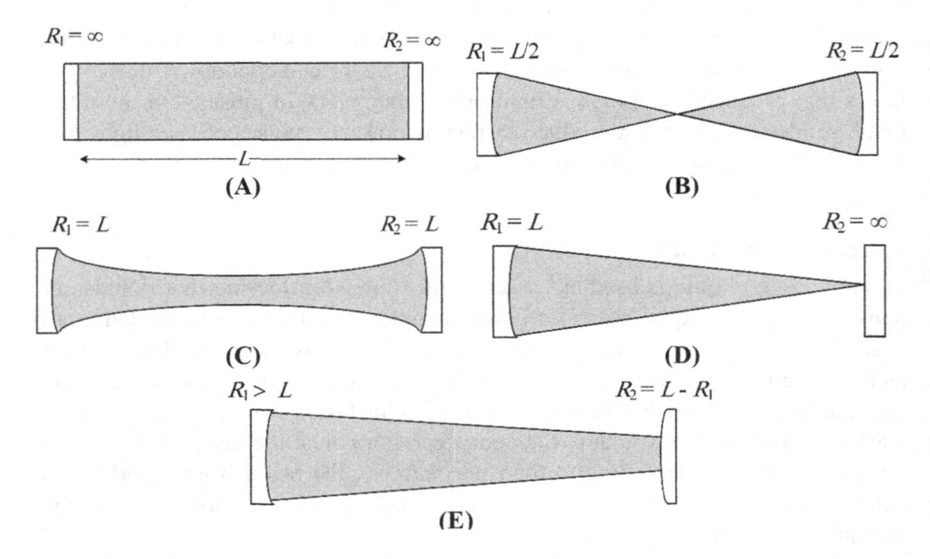

FIGURE 1.8 Types of optical resonators: (a) parallel-plane resonator, (b) spherical or concentric resonator, (c) concentric resonator, (d) hemispherical resonator, (e) concave-convex resonator.

The concave-convex resonator is also seen in part e of Figure 1.8. The focal radius of one of the lenses of this resonator is greater than the distance between the two lenses. The focal radius of the second lens is also equal to the difference between the distance between the two lenses and the focal radius of the first lens.

1.5 OUTPUT WAVES BASED ON TIME MODES

The output waves from the laser can be irradiated on the workpiece in two modes: continuous or pulsed. If the output beam from the laser is continuous and constant over time, the time mode is called a continuous process, and the laser is called a continuous wave mode (part a of Figure 1.9). The laser beam can be emitted over a repetitive and non-continuous time interval. Such a process is called laser with pulse beam mode (part b of Figure 1.9) [48]. In the following, each laser time mode will be introduced.

1.5.1 Continuous Waves or CW

Normally, the laser beam is continuously exited from the optical bag [14]. The output power of CW lasers has a wide range. Liquid, ion, gas lasers, and some solid lasers such as Nd:YAG and semiconductors are examples of continuous lasers [3,48]. This ranges from milliwatts, such as optical communication lasers or optical scanners, to several megawatts (up to 5 MW), such as lasers used in military industries [39].

1.5.2 Discrete Waves or DW

The output beam from the laser can be discrete. The output power of these lasers is much higher than that of CW lasers and can reach up to 1,015 W [39]. Some solid lasers, such as the ruby laser and Nd: Glass are two examples of discrete lasers [3]. According to the interval of wave repetition and the shape of the laser output, discrete waves can be divided into several types of femtosecond lasers: normal pulse [48], *Q*-switch [48,49], and mode locking [14,48,49]. Each is explained below.

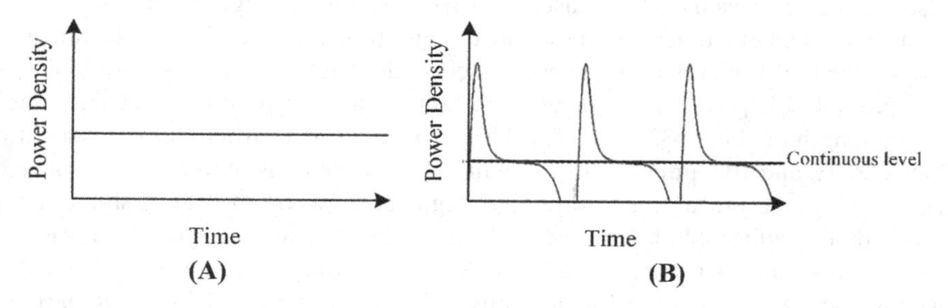

FIGURE 1.9 Output waves from the laser: (a) continuous and (b) pulsed.

1.5.2.1 Femtosecond Laser

When the separation interval of laser waves is 5 fs, it is called a femtosecond laser. Note that each femtosecond (fs) is equal to 10–15 seconds [14]. Some fluid lasers are examples of these lasers [50].

1.5.2.2 Normal Pulsed Lasers

When the dissociation interval of the wave output from the laser is in the range of picoseconds, nanoseconds, milliseconds, and microseconds, it is called a normal pulse laser. The examir laser is an example of this laser [51]. In the normal pulse process, the duration of the pulse is determined by changing the parameters of the flash lamp, such as the capacitor. Each primary pulse is applied with a maximum power that is approximately two to three times the average power of the pulse length. After the initial increase, the power decreases [48].

1.5.2.3 *Q*-Switch Lasers

By changing the mirrors in the resonator chamber, it is possible to change the properties of the laser beam pulse coming out of the laser. In *Q*-switch conditions, short and intense pulses are provided by changing the *Q* parameter of the cavity. The cavity *Q* parameter represents the cavity's ability to store radiant energy. When the value of *Q* is high, the energy is stored in the cavity, and a small amount of it is spent on the output laser beam. On the other hand, a low *Q* value indicates a laser beam with short pulses but with high energy. Therefore, *Q*-switch is defined in the sense of setting the value of the *Q* parameter with the aim of creating a beam with short and intense pulses. The *Q* parameter can be adjusted in various ways, for example, by rotating mirrors, electro-optics, auxo-optics, etc. With this type of lasers, pulses of several hundreds of kHz can be created [48].

1.6 TYPES OF LASERS

The world of lasers is so wide that its types can be divided into different points of view, such as the type of blower, active medium, optical resonator, time mode of the output beam, and wavelength of the output beam. The type of blower, active medium, and optical resonator were presented. All types of lasers can be classified according to each of these cases, which can be seen in Figure 1.10.

The wavelength range of the beam output from the laser is a wide range. According to this wavelength, types of lasers include infrared, visible, ultraviolet, and X-ray lasers [14]. For example, the helium-neon laser produces a beam with a wavelength of about 632.8 nm [36]. The wavelength of a carbon dioxide laser is between 9.6 and 10.6 μm [14]. This is while the wavelengths of fiber and excimer lasers are in the infrared and ultraviolet regions, respectively [36]. It should be noted that in adjustable lasers, the wavelength can be variable. For more information, you can refer to reference [52]. Also, depending on the wavelength and average power, lasers, according to Figure 1.11, can be considered in three parts: material processing, measurement, and communication.

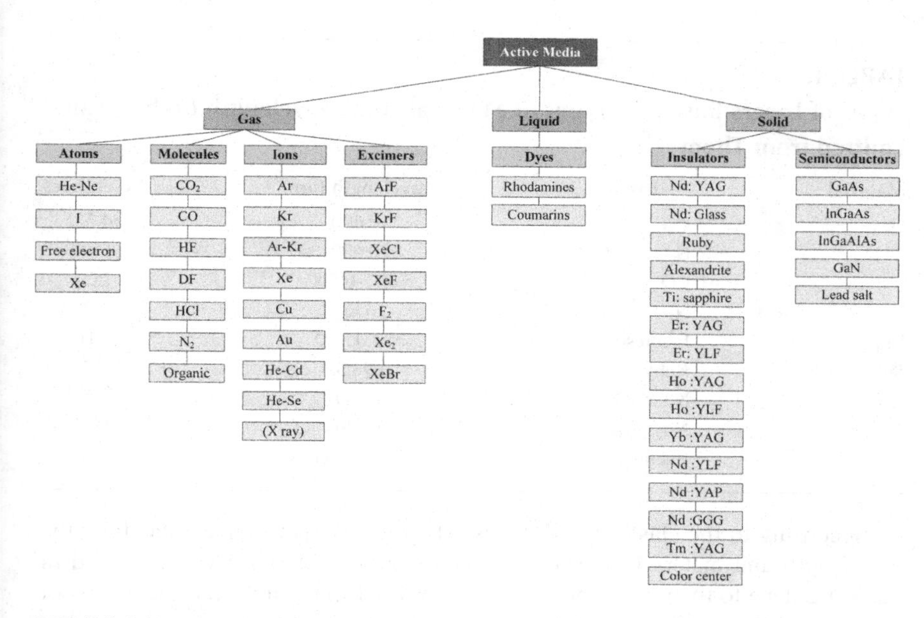

FIGURE 1.10 Classification of laser types based on the active medium.

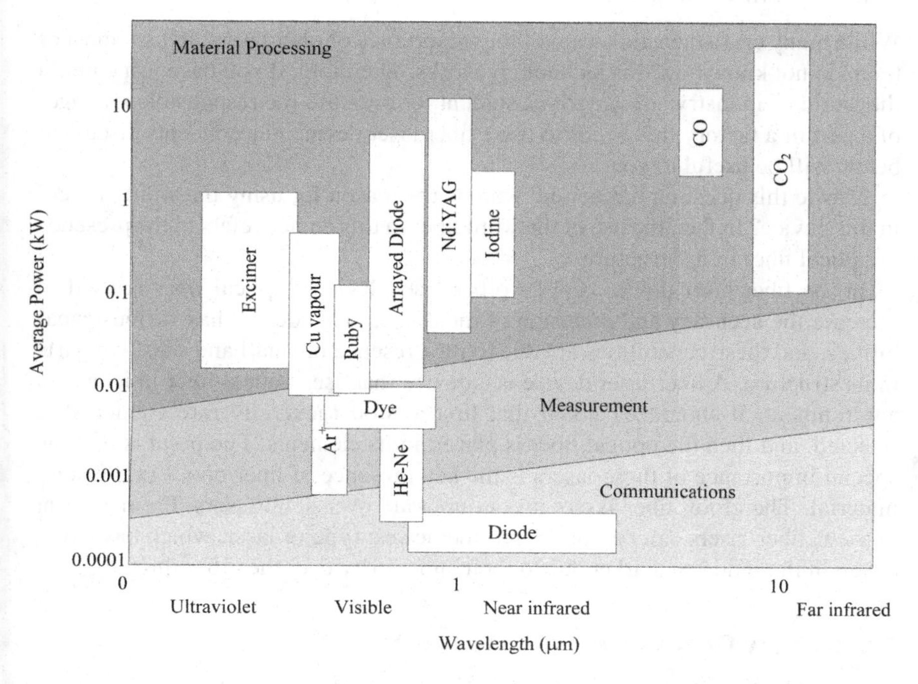

FIGURE 1.11 Classification of laser types based on wavelength and average power.

TABLE 1.2
Types of Lasers Based on Types of Material, and Wavelength of the Beam Emitted from Them

Material	Laser	Wavelength (μm)	References
Gas	CO_2	10,600	[4,35]
	N_2O	10,600	[52]
	CO	5,000–6,500	[52]
	N_2	336	[52]
Liquid	Fluorescence	300–1,100	[4]
Solid	Ruby	694.3	[52]
	Nd:YAG	1,060	[35]
	Nd: Glass	1,050	[35]
	Excimer	Ultraviolet	[35]

According to the classification of laser types, for a better understanding, the wavelength and material type of some commonly used lasers are presented in Table 1.2. Due to the wide application of fiber laser in the industry, the details of this type of laser will be presented below.

1.6.1 What Is Fiber Laser?

While many craftsmen are aware of the importance of fiber lasers, in fact, most of them do not know how this technology works. Therefore, if you have just entered the world of industry, are a curious student, or even are the responsible manager of a part of a factory that needs to use a fiber laser device, the concepts discussed below will be useful to you.

Maybe this question has arisen: what is the reason for using the word "Fiber" in this device? In fact, the use of the word fiber in this device refers to the presence of optical fiber in its structure.

In the fiber laser device, unlike other laser devices, optical fiber is used to increase the accuracy and precision of the device. This device has various capabilities, and these capabilities are due to the presence of small and sensitive parts in its structure. A fiber laser device is a device that uses optical fiber in its active environment. It should be noted that first a solid (generally rare elements) is selected, and then the optical fiber is placed in its elements. The point is that the special importance of these lasers is the key presence of fiber optics in the solid material. Therefore, fiber lasers are an upgrade over solid lasers. From a point of view, fiber lasers can be considered the newest type of laser, which has led to a leap in the industry market due to their superiority over the other three types.

1.6.2 How Does a Laser Fiber Machine Work?

As stated before, the optical fiber is applied as the main core of the active laser medium and is placed inside the elements of some rare elements (mostly Erbium). This is because the elements of these atoms have a high atomic energy level that allows a cheaper and less powerful energy source (usually a diode laser) to excite them well. Thus, the energy output will be very impressive. For example, by placing an optical fiber in erbium, the energy level that absorbed photons with a wavelength of 980 nm can absorb an equivalent level with a wavelength of 1,550 nm. This means that with an energy source capable of absorbing 980 nm energy, it is possible to produce a laser with a beam of 1,550 nm wavelength, which has more energy. Be careful that the fiber laser light beam is placed in the infrared region.

The laser production environment in the fiber laser device is a solid environment in which gray rare elements such as erbium, ytterbium, neodymium, diprosium, praseodymium, thulium, and holmium can be used to create laser radiation. In this way, energy enters the solid medium and causes the atoms to be excited. The atoms of the material in the active medium emit light from themselves, which is the same as laser radiation, but at this stage it has not yet become a focused beam and does not have enough energy to carry out the process. In order to create a strong and focused beam with high precision, the optical fiber in its path amplifies the beam and directs it to the lens. The lens focuses the rays at a very small point, and finally, a very strong beam is released from the device, which is performed with high precision at the focal point of the desired process. In short, fiber laser light is produced by a diode laser. Light is channeled and amplified within simple fiber optic cables. The amplified beams are aligned and straightened by the collimator when leaving the fiber cable. Then they will be focused and radiated by the lens.

1.6.3 Comparing Fiber Laser Versus CO_2 Laser?

In the following, for a better understanding, a comparison of 2.2–6 kW carbon dioxide laser and 2–4 kW fiber laser will be presented.

- Beam production using fiber laser is 200% more economical than traditional carbon dioxide lasers.
- Focusing the beam on the workpiece is easier with a fiber laser than with a carbon dioxide laser.
- Optical mirrors used for light transmission in fiber lasers are much cheaper than carbon dioxide lasers.
- Unlike the traditional carbon dioxide laser, there are no moving parts in the fiber laser beam production process. This has the advantage of reducing costs related to repair, maintenance, and moving parts.
- The electrical energy required by fiber lasers is far less than that of carbon dioxide lasers. In comparison, for the same process with the same accuracy, the energy required for a 3 kW fiber laser device is reported to be one third of that of 4 kW carbon dioxide laser device.
- If the metal workpiece is reflective, the carbon dioxide laser beam will be reflected. This is when the fiber laser light beam is focused on the shiny metal workpiece without reflection. In other words, the fiber laser device has no material limitations, and parts such as copper, brass, and aluminum that reflect light can also be cut with the laser device.
- While the light from the fiber laser passes through the glass workpiece, the light from the carbon dioxide laser cannot pass through the glass.

1.7 APPLICATION OF LASER MATERIAL PROCESSES IN INDUSTRY

Laser is a living topic in science that has occupied the minds of many researchers and scientists. The proof of this is the inventions that are available to mankind from moment to moment in this field. One of the broadest areas of laser application is laser material processing processes, which have various and important applications in various industries such as medicine, automotive, petrochemical, steel, aerospace, shipbuilding, etc. "Laser material processing" refers to a set of laser processes in which some properties of the workpiece are improved. This property can be connecting, cutting, additive manufacturing, increasing surface hardness, etc. In the following, some laser material processing will be introduced.

Laser material processing can be divided into marking, cutting, welding, coating, additive manufacturing, hard work, etc. The advantages of these laser methods over other traditional methods are safety, environmental friendliness, non-contact, very high accuracy, and high speed. Based on industrial applications in the field of materials, laser processes can be classified based on energy density, power density, and interference time, according to Figure 1.12.

According to Figure 1.12, according to the interference time, energy density, and power density, the laser beam is used in various laser material processing processes. The average energy density of laser cutting, melting, and hardening processes is approximately 10 J/mm^2. This is while the energy density of the key point welding and laser cladding processes is reported to be approximately 1,000 and 100 J/mm^2, respectively. Also, key point laser welding and laser machining require the highest and lowest laser power density in the process, respectively. From the point of view of interference time, the highest and lowest values are used for laser coating/laser hardening processes (approximately 10 seconds) and laser melting (approximately 0.001 seconds), respectively. It is worth noting that this comparison is actually very general and needs further investigation. Because all these processes belong to the family of laser material processing and laser is used to heat them, the physics of each process is very different from each other.

The "Laser Additive Manufacturing" or "LAM" process is a set of material processing processes with the help of a laser in which a layer of the desired material is created on the workpiece. This process can be done as a single layer or multiple layers with goals such as creating a coating or making a new piece. According to part a of Figure 1.13, in this process, energy is irradiated by the laser beam on the workpiece, and the molten pool is formed. The secondary material can be injected into the molten pool in the form of powder or wire. The secondary metal will melt after entering the molten pool. The presence of protective gas prevents the entry of pollutants into the environment; the molten material is slowly frozen, and a new layer is formed on the workpiece.

The process of "laser heat treatment" or "LHT" is another material processing process with the help of a laser, which is also referred to as "laser hard working" in some references. In this process, the laser is applied as a source of thermal energy with the aim of improving some surface quality parameters, such as hardness, on the workpiece. According to part b of Figure 1.13, in the LHT process,

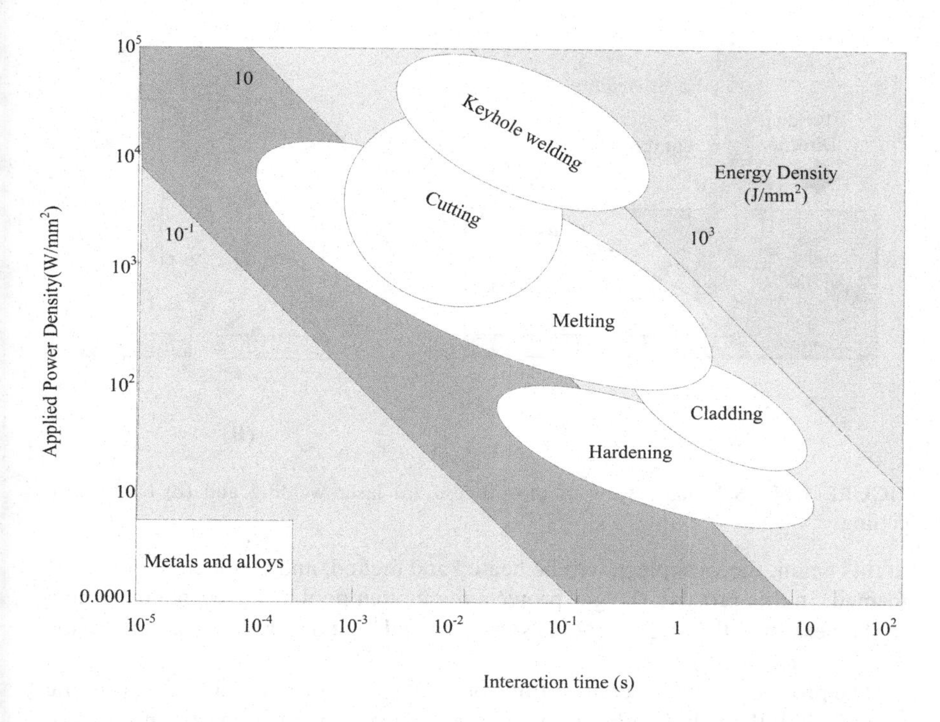

FIGURE 1.12 Classification of types of laser material processes based on energy density, power density, and interference time.

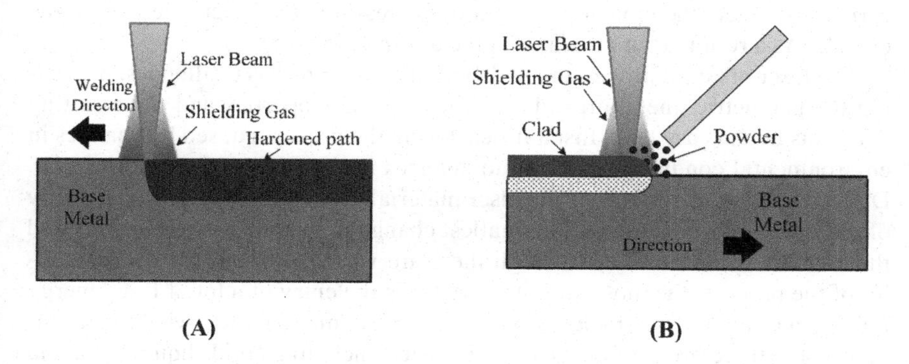

FIGURE 1.13 Schematic of the process: (a) laser hardening and (b) laser cladding.

a thin layer of the workpiece surface is heated and cooled at a high speed. As a result, a fine grain structure will be created in the workpiece, even for steels that have low hardenability.

The process of "laser beam welding" or "LBW" is a type of material processing process with the help of a laser, the purpose of which is the connection between two parts. According to part a of Figure 1.14, in this process, which is one of the melting processes, the laser beam is irradiated on the workpiece. Due

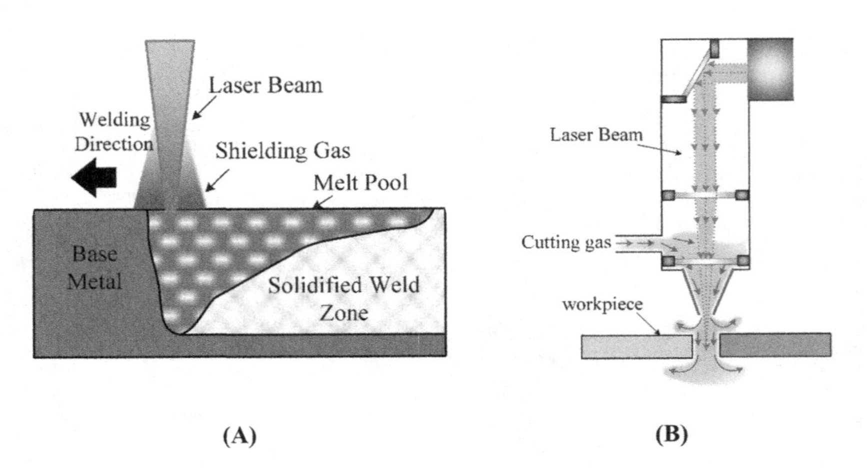

FIGURE 1.14 Schematic view of the process: (a) laser welding and (b) laser beam cutting.

to this beam, the workpiece will be heated and melted, and a molten pool will be formed. In this process, the gas protects the molten pool and prevents oxidation. In the next step, the molten pool is slowly cooled, and as a result, the connection will be established [4].

The process of "laser beam cutting" or "LBC" is a type of material processing process with the help of a laser, in which the purpose is to cut parts. In this process, a schematic view of which is shown in part b of Figure 1.14, the laser beam as an energy source leads to the melting of the workpiece. In the next step, the carrier gas enters the molten pool with high pressure and directs the melt to the outside. As a result, a cut is created at the desired location.

The laser material processing in different conditions has challenges; predicting the geometric, metallurgical, and mechanical properties and the resulting defects is not an easy task. Instabilities during the process caused by changes in environmental conditions can lead to changes in the structure of the material. Due to the high sensitivity of the laser material processing, the reproducibility of the test is questioned with the smallest change in the process parameters and the creation of instability. Although there are various reasons for the instability of the process, the most important factor is the entry of a lot of laser energy into a small area of the molten pool. Also, in the molten pool, which is a limited space, there are all four phases of matter, including solid, liquid, gas, and plasma. Therefore, the main challenge of laser material processing is its very high sensitivity to environmental conditions and process parameters. In other words, a small change in process parameters and environmental conditions may produce unacceptable results. For example, while the laser power of 1,900 W is considered a desirable parameter for the welding process, the laser power of 2,000 W will lead to significant and unmeasurable distortion in the process.

In general, laser material processing is a solidification process. For this reason, a combination of thermodynamic physical phenomena and kinetic reactions leads to the creation of metallurgical properties, mechanical properties, geometric properties, and various defects in the part. As a result, different efficiency will be obtained in the workpiece [53]. A view of this insight is shown in Figure 1.15.

According to Figure 1.15, equilibrium phases such as ferrite, austenite, etc. are extracted using thermodynamic information such as the maximum temperature of the molten pool and phase diagrams. The way of changing phases and morphology in thermal processes, which are a type of kinetic reaction, is obtained from the parameters of cooling rate and heating rate using time-temperature transfer diagrams and solidification diagrams. Metallurgical and crystallographic properties about the atomic structure and its arrangement in materials will be available by scanning electron microscopy [53]. By having crystallographic details or metallurgical properties, mechanical properties, geometric properties, and defects will be understood. These properties also indicate the efficiency of welding, cutting, or cladding resulting from laser material processing. Therefore, practically, by having the temperature details of the process, its efficiency will be understood. Although, according to Figure 1.15, the laser material processing seems simple, the understanding of the physics of each of the processes and their application is different. From a physical point of view, laser material processing can be classified into three categories: (1) thermal (without melting and evaporation), (2) melting (without evaporation), and (3) evaporation.

According to Figure 1.16, by choosing the laser power density and interference time, the temperature and state of the desired metal can be selected [54]. Due to the differences in behavior presented for each of the processes, each of the laser material processing steps will be carefully examined in each chapter of this book. In this regard, in the second chapter, the application and concepts governing the laser cutting process are introduced. In the third and fourth chapters, various types of welding methods and applications of the laser welding process will be presented, respectively. According to the mentioned contents, in the fifth chapter, the concepts governing the conventional laser welding process are introduced. In the sixth chapter, one of the latest

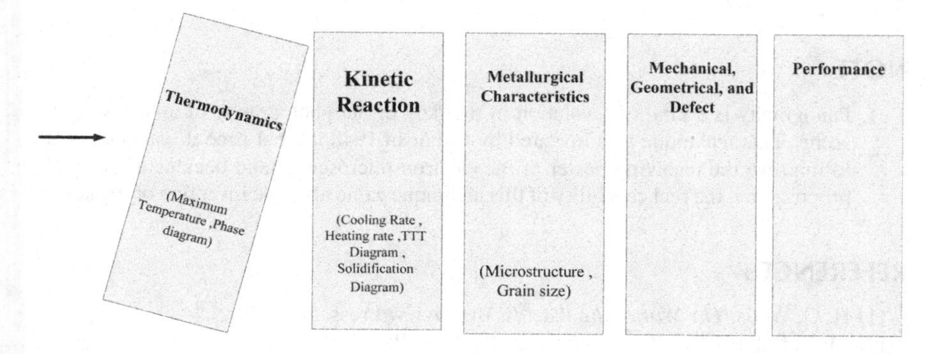

FIGURE 1.15 The relationship between thermodynamic properties, kinetic reaction, metallurgical, mechanical, geometrical characteristics, defect, and performance.

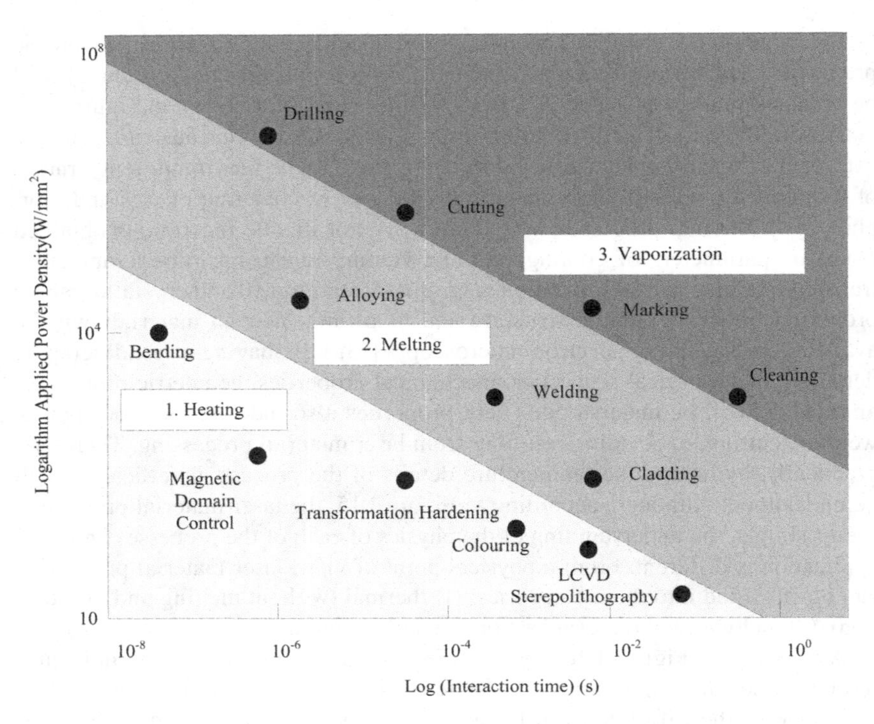

FIGURE 1.16 Classification of laser material processing based on three stages of heating, melting, and evaporation.

technologies of the day, laser wave welding, will be presented and carefully examined. In the seventh chapter, the laser coating process, including various methods and its applications, is introduced. Finally, in the eighth chapter, one of the most powerful areas of laser presence, i.e., additive manufacturing, will be presented and carefully examined.

NOTE

1. Panography is a kind of revolution in the field of 3D photography of an object or a scene. This technique was invented by Gabor in 1948 (at that time it was proposed to improve the resolving power of the electron microscope) and became a practical proposal, but the real capability of this technique came after the invention of the laser.

REFERENCES

[1] H. G. Wells, *The War of the Worlds*. Broadview Press, 2003.
[2] M. Bertolotti, *The History of the Laser*. CRC Press, 2004.
[3] C. W. Billings, *Lasers: New Technology of Light*. Universities Press, 1992.
[4] W. M. Steen and J. Mazumder, *Laser Material Processing*. Springer, 2010.
[5] W. Demtröder, *Laser Spectroscopy: Basic Concepts and Instrumentation*. Springer, 2013.

[6] A. Corney, *Atomic and Laser Spectroscopy*. Clarendon Press Oxford, 1978.
[7] G. D. Reid and K. Wynne, "Ultrafast laser technology and spectroscopy," *Encyclopedia of Analytical Chemistry*, vol. 35, no. 4, p. 8098, 2000.
[8] G. Nemova, *Laser Cooling: Fundamental Properties and Applications.* Jenny Stanford Publishing, 2016.
[9] V. Vuletić and S. Chu, "Laser cooling of atoms, ions, or molecules by coherent scattering," *Physical Review Letters*, vol. 84, no. 17, p. 3787, 2000.
[10] L. Rieg, V. Wichmann, M. Rutzinger, R. Sailer, T. Geist, and J. Stötter, "Data infrastructure for multitemporal airborne LiDAR point cloud analysis-Examples from physical geography in high mountain environments," *Computers, Environment and Urban Systems*, vol. 45, pp. 137–146, 2014.
[11] D. Cunningham, S. Grebby, K. Tansey, A. Gosar, and V. Kastelic, "Application of airborne LiDAR to mapping seismogenic faults in forested mountainous terrain, southeastern Alps, Slovenia," *Geophysical Research Letters*, vol. 33, no. 20, p. L20308, 2006.
[12] C. Weitkamp, *Lidar: Range-Resolved Optical Remote Sensing of the Atmosphere.* Springer, 2006.
[13] J. Hecht, "Lidar for self-driving cars," *Optics and Photonics News*, vol. 29, no. 1, pp. 26–33, 2018.
[14] F. R. Karl, *Basics of Laser Physics For Students of Science and Engineering.* Spinger, 2017.
[15] A. L. Schawlow and C. H. Townes, "Infrared and optical masers," *Physical Review*, vol. 112, no. 6, p. 1940, 1958.
[16] A. F. Seyedeh Fatemeh Nabavi and M. Hossein Farshidianfar, "Fundamental of laser and its application in industries," *Mechanical Engineering*, vol. 28, no. 4, pp. 61–71, 2020.
[17] W. E. Lamb Jr and R. C. Retherford, "Fine structure of the hydrogen atom by a microwave method," *Physical Review*, vol. 72, no. 3, p. 241, 1947.
[18] R. H. Dicke, "Coherence in spontaneous radiation processes," *Physical Review*, vol. 93, no. 1, p. 99, 1954.
[19] N. Bloembergen, "Proposal for a new type solid state maser," *Physical Review*, vol. 104, no. 2, p. 324, 1956.
[20] A. L. Schawlow and C. H. Townes, "Masers and maser communications system," Google Patents, 1960.
[21] R. Ladenburg, "Research on the anomalous dispersion of gases," *Physikalische Zeitschrift*, vol. 48, pp. 15–25, 1928.
[22] T. H. Maiman, "Stimulated optical radiation in ruby," *Nature*, vol. 187, pp. 493–494, 1960.
[23] P. Sorokin and M. Stevenson, "Stimulated infrared emission from trivalent uranium," *Physical Review Letters*, vol. 5, no. 12, p. 557, 1960.
[24] A. Javan, W. R. Bennett Jr, and D. R. Herriott, "Population inversion and continuous optical maser oscillation in a gas discharge containing a He-Ne mixture," *Physical Review Letters*, vol. 6, no. 3, p. 106, 1961.
[25] R. N. Hall, G. E. Fenner, J. Kingsley, T. Soltys, and R. Carlson, "Coherent light emission from GaAs junctions," *Physical Review Letters*, vol. 9, no. 9, p. 366, 1962.
[26] J. Geusic, H. Marcos, and L. Van Uitert, "Laser oscillations in Nd-doped yttrium aluminum, yttrium gallium and gadolinium garnets," *Applied Physics Letters*, vol. 4, no. 10, pp. 182–184, 1964.
[27] C. K. N. Patel, "Continuous-wave laser action on vibrational-rotational transitions of CO_2," *Physical Review*, vol. 136, no. 5A, p. A1187, 1964.
[28] W. B. Bridges, "Laser oscillation in singly ionized argon in the visible spectrum," *Applied Physics Letters*, vol. 4, no. 7, pp. 128–130, 1964.

[29] J. V. Kasper and G. C. Pimentel, "HCl chemical laser," *Physical Review Letters*, vol. 14, no. 10, p. 352, 1965.
[30] W. T. Silfvast, G. R. Fowles, and B. Hopkins, "Laser action in singly ionized Ge, Sn, Pb, In, Cd and Zn," *Applied Physics Letters*, vol. 8, no. 12, pp. 318–319, 1966.
[31] M. L. Geiges, "History of lasers in dermatology," In: Bogdan Allemann and D.J. Goldberg (eds.) *Basics in Dermatological Laser Applications*. Karger Publishers, vol. 42, pp. 1–6, 2011.
[32] R. George, "Laser in dentistry: Review," *International Journal of Clinical Dentistry*, vol. 1, no. 1, pp. 13–19, 2009.
[33] S. T. Holmström, U. Baran, and H. Urey, "MEMS laser scanners: A review," *Journal of Microelectromechanical Systems*, vol. 23, no. 2, pp. 259–275, 2014.
[34] G. D. Gautam and A. K. Pandey, "Pulsed Nd:YAG laser beam drilling: A review," *Optics & Laser Technology*, vol. 100, pp. 183–215, 2018.
[35] W. Wang, B. Zhou, S. Xu, Z. Yang, and Q. Zhang, "Recent advances in soft optical glass fiber and fiber lasers," *Progress in Materials Science*, vol. 101, pp. 90–171, 2019.
[36] C. B. Hitz, J. J. Ewing, and J. Hecht, *Introduction to Laser Technology*. John Wiley & Sons, 2012.
[37] H. M. Presby, "Near 100% efficient fiber microlenses," In: *Optical Fiber Communication Conference*. Optica Publishing Group, p. PD24, 1992.
[38] N. Hodgson and H. Weber, *Optical Resonators: Fundamentals, Advanced Concepts, Applications*. Springer, 2005.
[39] O. Svelto and D. C. Hanna, *Principles of Lasers*. Springer, 2010.
[40] A. E. Siegman, *Lasers*. University Science Books, 1986.
[41] R. S. Quimby, *Photonics and Lasers: An Introduction*. John Wiley & Sons, 2006.
[42] C. Yeh, *Applied Photonics*. Elsevier Science, 1994.
[43] M. Eichhorn, *Laser Physics: From Principles to Practical Work in the Lab*. Springer, 2014.
[44] A. Mendelsohn, R. Normandin, S. Harris, and J. Young, "A microwave-pumped XeCl* laser," *Applied Physics Letters*, vol. 38, no. 8, pp. 603–605, 1981.
[45] A. E. Siegman, "Lasers university science books," *Mill Valley, CA*, vol. 37, no. 208, p. 169, 1986.
[46] T. Hansch, M. Pernier, and A. Schawlow, "Laser action of dyes in gelatin," *IEEE Journal of Quantum Electronics*, vol. 7, no. 1, pp. 45–46, 1971.
[47] M. J. Beesley, *Lasers and their Applications*. Taylor & Francis Group, 1976.
[48] N. B. Dahotre and S. Harimkar, *Laser Fabrication and Machining Of Materials*. Springer, 2008.
[49] J. F. Ready, *LIA handbook of Laser materials processing*, Laser Institute of America Magnolia, 2001. ISBN: 0912035153.
[50] J. S. Hoppius, S. Maragkaki, A. Kanitz, P. Gregorčič, and E. L. Gurevich, "Optimization of femtosecond laser processing in liquids," *Applied Surface Science*, vol. 467, pp. 255–260, 2019.
[51] S. N. Joffe and Y. Oguro, *Advances in Nd:YAG Laser Surgery*. Springer, 2012.
[52] F. Duarte, "Tunable laser microscopy," In: Duarte F. J. (ed.), *Tunable Laser Applications*. CRC Press, pp. 315–328, 2016.
[53] C. Gobbi, *Low Cost Thermal Imaging System for Welding Applications*. University of Waterloo, 2016.
[54] J. Dutta Majumdar and I. Manna, "Laser processing of materials," *Sadhana*, vol. 28, pp. 495–562, 2003.

2 Laser Beam Cutting Process
Phenomena and Applications

2.1 INTRODUCTION TO CUTTING

The concept of utilizing light for cutting traces back to ancient times, when early humans attempted to focus sunlight onto paper surfaces using glass to achieve a cutting effect. It is important to acknowledge that concentrated sunlight causes the paper to burn, thereby creating a cut (Figure 2.1). At first glance, this primitive approach appears attractive due to its simplicity. The notion of harnessing and concentrating light, which would otherwise disperse in the atmosphere, to cut paper gradually evolved into a gateway for addressing industrial challenges as technology advanced. Over time, this technology experienced remarkable progress, eventually enabling the cutting of steel with thicknesses measured

FIGURE 2.1 Burning paper with sunlight.

DOI: 10.1201/9781003492191-2

in centimeters. The evolution of this technique showcases the immense growth and potential of laser cutting, providing the industry with unprecedented opportunities for precision and efficiency.

The laser cutting process holds significant importance and is recognized as one of the most prevalent cutting methods in contemporary times. While the captivating nature of laser cutting alone has not been sufficient to capture the attention of the industry, its widespread adoption can be attributed to its remarkable attributes of speed, precision, and high quality. With the rapid rise of laser technology in the global market, experts and professionals alike consider laser cutting to be an established technology.

Statistics demonstrate the uniqueness and pioneering nature of the laser cutting process within the laser industrial sector. A study conducted in 2004 revealed that laser cutting machines accounted for 10% of machinery sales in North America [1]. Furthermore, laser cutting surpasses other laser-assisted material processing techniques, as highlighted by a study in 2010 indicating that 80% of lasers employed in the Japanese industry were utilized for laser cutting purposes [2]. Additionally, a study in 2017 projected that the global laser cutting market would reach a value of 5.7 billion dollars by 2022, showcasing a growth rate of 9.3% compared to 2016. In light of these statistics, it is evident that the laser cutting process meets a pressing demand. In the subsequent section, the applications of laser cutting will be comprehensively presented.

2.2 APPLICATION OF THE LASER CUTTING PROCESS

One of the most widely applied uses of laser cutting technology is in the fabrication of metal parts. This versatile process finds application across various metals, including steel, tungsten, nickel, brass, and aluminum, among others. Irrespective of the specific industry, metals or metal combinations are typically encountered in the vicinity of laser cutting operations. The demand for precise and efficient cutting of these materials has propelled the adoption of laser cutting in these domains. With the ability to effortlessly cut desired pieces of any shape with exceptional precision, laser cutting proves capable of handling thicknesses ranging from 0.5 to 30 mm. These pieces may encompass components as diverse as car bodies or mobile phone parts. Thus, laser cutting has emerged as a powerful tool within production processes.

The adoption of laser cutting spans numerous large-scale industries, including aviation, automotive, electronics, semiconductors, and medicine. This widespread adoption can be attributed to the substantial advantages conferred by laser cutting, which have revolutionized industries in terms of quality and reliability compared to conventional methods. Notably, the laser cutting process offers features such as high speed and obviates the need for extensive post-processing, rendering it particularly suited for cutting thin knife edges. To elucidate the significance of these disparities and highlight the distinctiveness of laser cutting, several industrial applications of this process will be detailed below.

2.2.1 Aerospace

The unique and demanding conditions to which aerial structures are exposed underscore the significance and sensitivity of their components. The laser cutting process finds extensive application in air structures such as airplanes, defense industries, and spaceships. As air travel remains one of the most efficient means of transportation, offering global accessibility and time savings, it has garnered significant attention in recent years from both the international community and industries.

The aerospace industry necessitates the utmost precision, given its critical nature. The demand for near-zero production tolerances stems from the industry's emphasis on accuracy. Moreover, considering the high cost of components, it is imperative to employ production methods that minimize the occurrence of defective parts, human error, and time limitations. In response to these requirements, the aerospace sector has increasingly adopted precise cutting processes. In the aircraft industry, laser cutting is employed extensively in the fabrication of various components, including end parts of the body, wings, and moving parts. These components primarily consist of materials such as stainless steel, titanium, nickel alloy, iron, inconel, aluminum, and tantalum, among others. It is noteworthy that carbon dioxide and ultraviolet laser devices are commonly utilized for laser cutting in the aerospace industry.

The integration of laser cutting technology in the aerospace industry serves to enhance the overall quality and efficiency of production processes. The precise and controlled nature of laser cutting aligns with the industry's stringent requirements, contributing to the manufacturing of high-performance aerospace structures.

It is worth highlighting that the utilization of laser cutting in these cases aims to minimize waste and damage to aircraft parts while ensuring high speed and quality production, given the inherent sensitivity of such components. For instance, the inner aluminum part of the cockpit in the nose section of the Boeing 777 is manufactured using laser cutting techniques. Among the most critical components within this section is the Electronic Flight Instrument System (EFIS), which, along with other control elements in the internal nose area, is produced using laser cutting technology. These components are vital for the aircraft's functionality and safety.

The significance of laser-cut parts in the aerospace industry is evident from the market demand they generate. To exemplify the practicality and market value of such components, it is noteworthy that the sale of these laser-cut parts in the United States alone reached approximately $1.139 million in 2021. The application of laser cutting technology in the aerospace sector demonstrates its effectiveness in achieving precise and high-quality production while catering to the specific requirements and sensitivity of aircraft parts. By reducing waste, minimizing damage, and ensuring efficient manufacturing, laser cutting plays a crucial role in enhancing the overall performance and safety of aircraft structures.

Another notable application of the laser cutting process in the aviation industry is evident in the production of airplane rudder pedal gears. These pedals play a critical role in controlling the aircraft's rudder. While the thickness and material composition may vary depending on the specific aircraft model, they are typically constructed from titanium 6061-T6 and possess a thickness of approximately 4 mm.

The utilization of laser cutting technology in fabricating these pedal gears offers numerous advantages, including precise cutting and shaping of the titanium material. Laser cutting ensures high accuracy and allows for intricate designs to be implemented, thus facilitating the optimal functionality and performance of the rudder pedal gears. To explore further applications of laser cutting in the aviation industry, it is recommended to refer to studies [3–5]. These studies provide valuable insights into the diverse and significant uses of laser cutting technology within the aerospace sector.

2.2.2 Automotive Industry

The automotive industry holds a prominent position in the global market, with staggering statistics highlighting its significance. In 2018 alone, approximately 80 million cars were sold worldwide, illustrating the immense scale of the industry. To put this into perspective, it equates to an astounding average of 220,000 cars sold each day. With such a high volume of cars being produced and sold, it becomes imperative that the parts and components are manufactured with precise accuracy to ensure optimal efficiency and performance. The quality of these vehicles must be upheld, as they consist of an average of 30,000 individual parts. Thus, if the automotive industry operates at its full capacity, approximately 6,600,000,000 parts need to be manufactured daily, meeting stringent quality and precision standards to ensure the safety and well-being of passengers.

The close relationship between automotive safety, market demand, and production volume underscores the paramount importance of reliability and efficiency within the manufacturing process. In this context, laser cutting assumes a highly valuable role in the automotive industry. The capital investment involved in producing such an enormous number of parts necessitates the assurance that laser cutting technology readily provides. The precision and versatility offered by laser cutting play a crucial role in meeting the stringent requirements of the automotive industry, enabling the fabrication of high-quality components that contribute to the overall performance, safety, and reliability of vehicles. The pivotal role of laser cutting in the automotive industry stems from its ability to deliver the necessary precision, speed, and quality to support the demanding production requirements. By ensuring the manufacturing of precise parts with superior accuracy, laser cutting technology contributes significantly to the success and sustainability of the automotive sector.

The laser cutting process finds extensive application in the fabrication of car body parts, which represent a significant domain within the realm of laser cutting.

The evolution of car design over the past few decades has led to the production of smaller and more intricate car body components. In response to this demand, the automotive industry has turned to advanced laser cutting technologies, such as fiber laser cutting and solid laser cutting devices, to achieve the precision and complexity required for various car parts.

These laser cutting technologies have been effectively employed in the production of car components such as electronic boards, gear shift levers, mirrors, panel buttons, hand controls, air filters, car body frames, license plates, crankshafts, wiper blades, engine blocks, bearings, small screws and nuts, and fuse boxes, among others. The versatility and precision of laser cutting enable the manufacturing of these intricate parts with high accuracy and efficiency.

The utilization of laser cutting in car body parts not only ensures the precise shaping and cutting of various materials but also contributes to the overall quality and functionality of the vehicles. Laser cutting technology plays a vital role in meeting the demands of modern car design, enabling the production of smaller and more intricate components that enhance the aesthetics, performance, and overall user experience of the vehicle.

By incorporating fiber laser cutting and solid laser cutting devices in the manufacturing process, the automotive industry can achieve the necessary precision and complexity required for the intricate car body parts of today's vehicles.

Laser cutting technology finds significant application in the fabrication of car body parts. For instance, Volkswagen utilizes a CO_2 laser cutting machine with specific parameters to perform the cutting process of the production body. The cutting process involves a beam diameter of 0.1 mm, a speed of 4.5 m/min, and a laser power of 400 W [6]. This application showcases the precision and effectiveness of laser cutting in the automotive industry.

Another study focused on investigating the laser cutting of a miniature rib made of stainless steel 304. The rib had an external diameter of 9.04 mm and a thickness of 4.5 mm. Laser cutting parameters such as laser power ranging from 1,500 to 2,500 W, cutting speed between 1 and 3 m/min, focal position varying from –1.5 to –3.5, gas pressure of 10–16 bar, nozzle diameter of 1.7 mm, and a nozzle-to-workpiece distance of 1 mm were examined. The cutting process was conducted using a CO_2 laser machine with a maximum power of 3.2 kW [7]. The study observed that high laser power, low cutting speed, medium focal position, and appropriate control of these parameters contributed to achieving low surface roughness and precise gear teeth.

It is noteworthy that miniature gears, similar to those used in the automotive industry, also find application in various components requiring power transmission and motion, such as miniature pumps, engineering tools, and medical instruments. The laser cutting process offers an efficient means of fabricating these intricate gear components, with specific laser parameters tailored to achieve desired outcomes.

These studies exemplify the diverse applications of laser cutting technology in the automotive industry, demonstrating its effectiveness in cutting car body parts, miniature ribs, and gears. By employing laser cutting techniques with careful control of parameters, manufacturers can achieve precise cuts, low surface roughness, and high-quality components for use in automobiles and other related industries.

Laser cutting technology plays a significant role in the manufacturing of hydroformed parts in the automotive industry. The complexity of the profile geometry of tubes and pipes used in hydroformed components necessitates the use of 3D laser cutting. In the case of the Lamborghini LP 770-4, a high-performance car renowned for its 770 horsepower, all the hydroformed parts are laser cut to ensure optimal efficiency. Although the use of laser cutting contributes to the higher cost of manufacturing these luxury cars, it is essential to meet the stringent standards set for such vehicles.

Apart from hydroformed parts, laser cutting is also employed in various other components of automobiles. Airbags, floor coverings, select car washers, and luxury car accessories utilize laser cutting technology. Further details regarding the specific applications of laser cutting in these components are beyond the scope of this discussion but can be explored in-depth by curious and enthusiastic readers. It is worth mentioning that CO_2 laser cutting machines are commonly used for airbag production and car accessories, while fiber laser cutting machines find extensive use in most other car parts. For a comprehensive understanding of laser cutting in the automotive industry, readers are encouraged to refer to the cited research study. This study provides further insights and analysis into the specific applications, techniques, and advancements of laser cutting technology in the automotive sector.

2.2.3 Electronic Industry

In the preceding sections, the applications of laser cutting in prominent industries like aerospace and automotive are explored. Similarly, laser cutting plays a pivotal role in the electronics industry, where precision and high-quality cutting are paramount. Behind the scenes of electronic device manufacturing and production, laser cutting stands as a powerful tool. To shed light on this application and elucidate its scope, we will now present some examples in this field.

In the electronics industry, laser cutting is utilized for various purposes, such as the fabrication of printed circuit boards (PCBs) and the production of intricate electronic components. Laser cutting enables the precise shaping and resizing of PCBs, allowing for the integration of complex circuitry on small and densely packed boards. This technology ensures the accuracy required for the manufacturing of modern electronic devices.

Laser cutting plays a crucial role in the manufacturing of intricate electronic components, particularly in the realm of microelectromechanical systems and microfluidic devices. These components are characterized by intricate patterns, channels, and structures at microscopic scales. Laser cutting technology provides a precise and efficient means to fabricate these complex features, contributing significantly to the development of miniaturized and highly functional electronic devices.

The ability of laser cutting to operate at a microscopic level enables the production of intricate geometries with high precision. This is particularly advantageous in the creation of microelectromechanical systems, where tiny mechanical elements and sensors need to be manufactured with utmost accuracy. Additionally, in the field of microfluidics, laser cutting is employed to create microchannels and intricate fluidic pathways in devices designed for tasks such as lab-on-a-chip applications. The non-contact nature of laser cutting further enhances its suitability for working with delicate materials commonly found in electronic components. As technology continues to advance, laser cutting remains a key enabling technology for the fabrication of intricate and miniaturized electronic devices, contributing to advancements in various industries, including electronics, healthcare, and telecommunications.

Furthermore, laser cutting is employed in the semiconductor industry for wafer dicing and microchip singulation. Laser cutting ensures clean and precise cuts, minimizing the risk of damage to delicate semiconductor materials and enhancing overall yield and efficiency in the production process. These examples illustrate the significant role of laser cutting in the electronics industry. By providing the necessary accuracy and quality in cutting processes, laser technology enables the production of advanced electronic devices and components. The continuous advancements in laser cutting techniques and systems further contribute to the progress of the electronics industry.

As anticipated, the electronics industry is vast and continuously expanding around the world. With the rapid growth of technology and the digital realm, the production of electronic components and products has become unstoppable. According to statistics from 2019, the annual consumption of electronic products exhibited a staggering growth rate of approximately 5%. These products encompass a wide range, including smartphones, tablets, televisions, and their accessories. It has been projected that the market value of these products will reach a remarkable $838.35 billion by 2020. When observing electronic products closely, it becomes evident that their dimensions and sizes are gradually decreasing over time. It is important to note that the combination of an escalating production growth rate, the need for product advancements, and the shrinking form factor has driven the industry towards adopting laser cutting in manufacturing processes.

In the pursuit of enhanced accuracy, speed, and quality, the laser cutting process is extensively employed in various components of electronic devices, particularly in the manufacturing of mobile phones. Prominent companies such as Samsung and Apple are notable examples of industry leaders that employ laser cutting technology to shape the bodies of their mobile phones. Laser cutting is employed in creating precise cuts and contours in the plastic and metal bodies of these devices. Additionally, laser cutting plays a crucial role in fabricating components within the mobile phone, including the camera lens, screen, answer key, and electronic chips. The presence of laser cutting technology in the production of mobile phones showcases its significant contribution to the manufacturing process.

By utilizing laser cutting technology, manufacturers in the electronics industry can achieve superior precision, faster production speeds, and high-quality results. Laser cutting has become an integral part of the production process, enabling the creation of smaller and more sophisticated electronic devices. As technology continues to advance, laser cutting will likely play an even more prominent role in the evolution of the electronics industry.

2.2.4 Medical

The field of medicine holds paramount importance in both material and spiritual aspects of human life, as it is intimately connected to the well-being and preservation of human life itself. Recognizing this significance, laser cutting technology has made its way into the medical industry. Laser cutting is extensively utilized in the production of medical devices and equipment, many of which are intricately small in size. It is crucial to emphasize that the medical field is concerned with human lives, and therefore, the quality and reliability of these devices should be impeccable from the outset and throughout the production process. This section will delve into the role of laser cutting in ensuring the reliability, quality, and efficacy of medical devices, as well as shed light on the reasons and methods by which laser cutting has entered the challenging and delicate profession of medicine.

In addition to medical instruments, laser cutting also finds applications in therapeutic methods and surgical procedures. The market value of the laser technology field in 2019 was estimated to be around $338 billion, and this value is expected to grow further with advancements in methods and devices. As the industry strives for progress, it also seeks ways to reduce costs and develop processes that meet the needs of both healthcare providers and patients without compromising on scientific principles. While the primary goal of the medical field is to preserve human life rather than generate profit, cost-effectiveness must be taken into consideration. Affordable healthcare and medical equipment should be accessible to individuals across all socioeconomic backgrounds.

One of the most critical applications of laser cutting technology lies in the construction of stents for angioplasty procedures. A stent, often described as a small metal mesh tube or spring, is permanently inserted into narrowed arteries to facilitate their dilation. When the coronary artery, responsible for supplying blood to the heart muscle, becomes narrowed due to plaque accumulation, blood flow to the heart muscle is compromised, leading to chest pain. In more severe cases, the formation of a blood clot within the artery can completely block blood flow, resulting in a heart attack.

The precise construction of stents using laser cutting techniques plays a vital role in ensuring their efficacy and longevity. Laser cutting allows for the creation of intricate and precisely designed stents that can be safely implanted within the arteries, restoring proper blood flow and mitigating the risk of further complications. The utilization of laser cutting technology in this context demonstrates its indispensable contribution to the medical field and its commitment to improving patient outcomes.

In summary, laser cutting has made significant inroads into the field of medicine, encompassing the production of medical devices, therapeutic approaches, and surgical interventions. The integration of laser cutting technology is driven by the pursuit of quality, reliability, and cost-effective healthcare solutions. Furthermore, laser cutting plays a pivotal role in the construction of stents used in angioplasty procedures, exemplifying its vital contribution to cardiovascular medicine and patient care [8].

To alleviate a narrowed artery, physicians often employ a procedure known as percutaneous coronary intervention (PCI) or angioplasty. During this process, a catheter, which is a tube with a balloon at its tip, is inserted into the affected artery and directed towards the site of the blockage. The series of steps involved in the procedure are illustrated in Figure 2.2, divided into segments A, B, C, and D.

In the initial step (part a of Figure 2.2), the catheter is carefully threaded into the artery until it reaches the narrowed region. Once the catheter is appropriately positioned, the balloon is inflated (part b of Figure 2.2). The inflation of the balloon exerts pressure on the plaque, compressing it and expanding the constricted area of the artery. This dilation process helps restore normal blood flow.

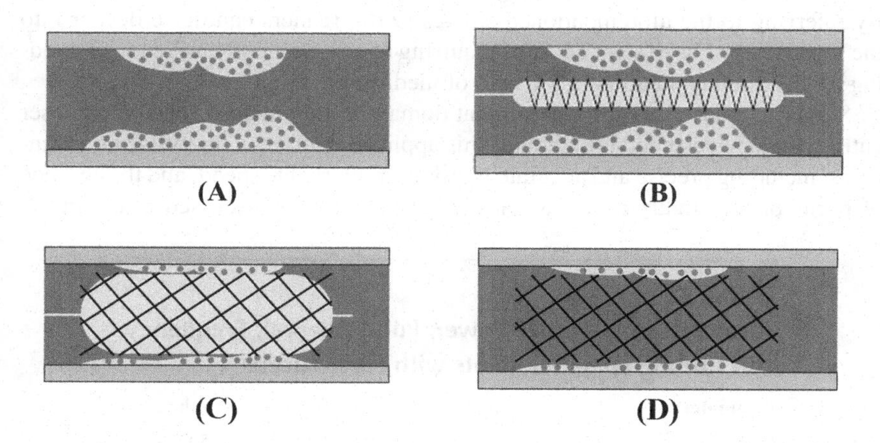

FIGURE 2.2 Stent placement steps in the artery: (a) vein with deposits, (b) stent insertion with balloon, (c) stent opening with balloon inflation, and (d) balloon exit and stent connection to the artery.

Following the successful dilation of the artery, the balloon is deflated (part c of Figure 2.2). Subsequently, the catheter is gently withdrawn from the artery, completing the procedure (part d of Figure 2.2). This allows the artery to resume its regular function, promoting improved circulation and reducing the risk of complications.

The schematic representation in Figure 2.2 provides a visual guide to the various stages of the percutaneous coronary intervention (PCI) or angioplasty procedure. It serves as a valuable reference for understanding the sequence of events involved in opening a narrowed artery and highlights the critical role of this procedure in the field of cardiovascular medicine [8].

According to a 2010 report, the annual placement of stents in patients' bodies exceeds three million. This statistic underscores the severe impact of coronary artery disease, which remains one of the most fatal illnesses worldwide. The significance of stents, however, extends beyond their application in arteries, as they are also utilized in the urinary tract and esophagus. Laser cutting has emerged as the simplest, fastest, and highest-quality method for manufacturing stents. For more detailed insights into laser stent cutting, further investigations can be found in studies [9,10].

In relation to the production of stents, Table 2.1 provides a compilation of fiber laser parameters employed in laser cutting [11]. This table serves as a valuable resource for comprehending the characteristics and specifications of the fiber lasers utilized in stent manufacturing. By leveraging laser cutting technology, medical professionals can produce stents with exceptional precision, ensuring their effectiveness in treating arterial blockages and other related conditions.

The widespread adoption of laser cutting techniques has revolutionized the stent production process, facilitating enhanced efficiency, reliability, and quality. By referring to the aforementioned studies [9,10], readers can delve deeper into the intricacies of laser stent cutting, acquiring a more comprehensive understanding of this vital application in the field of medicine.

Surgery stands as another prominent domain in the medical field where laser cutting finds widespread utilization. This application offers a multitude of advantages, including precise and accurate incisions, remarkable speed, and the absence of burns or heat-induced damage at the cutting site. As discussed in the initial

TABLE 2.1
Parameters of Average Power, Pulse Interval, Frequency and Cutting Speed of Stents with Fiber Lasers

Parameter	Values
Mean power	20–50 W
Pulse duration	0.05–0.15 ms
Frequency	0.5–5 kHz
Cutting speed	4–33 ms

chapters, laser cutting has made previously unthinkable surgical procedures feasible. Notably, the realm of eye surgeries exemplifies the remarkable potential of laser cutting devices. These advancements have enabled the execution of highly delicate procedures with enhanced precision and safety.

It is worth emphasizing that the scope of laser cutting in medicine extends beyond eye surgeries. The continuous development and exploration of this technology in surgical interventions have yielded remarkable outcomes. The application of laser cutting continues to expand, with numerous innovative procedures and advancements continually emerging. The readers, with their discerning taste, are encouraged to delve deeper into this topic to explore the comprehensive range of applications and advancements in the field.

2.2.5 Entertainment

In addition to its prominent role in various serious industries, laser cutting has also gained recognition in the entertainment world. One notable application is the creation of airplane models, which has captured the interest of children. An intriguing example is the aluminum and steel body of a 3D model developed by the Lexus team, featuring intricate Japanese origami-inspired patterns. This model possesses realistic mobility akin to that of an actual car, providing an immersive experience for enthusiasts.

Furthermore, laser cutting has made its way into the realm of art. Artists have embraced this technology to explore new horizons in jewelry design. Rings, earrings, necklaces, and chains are just a few examples of jewelry items that can be intricately crafted using laser cutting machines. This integration of laser cutting into the artistic domain has opened up a world of possibilities for designers and artisans to unleash their creativity.

The proliferation of laser cutting applications in entertainment, art, and jewelry highlights its wide-ranging impact and versatility. To ensure the continued success and advancement of laser cutting, extensive research and studies have been conducted in the fields of science and industry. Notably, numerous investigations have focused on exploring different types of lasers, pushing the boundaries of what laser cutting technology can achieve [12–14].

2.2.6 Heat Exchanger Systems

The laser cutting process finds diverse applications in heat transfer systems, particularly in the realm of plate heat exchangers. Plate heat exchangers have garnered considerable attention in recent years due to their numerous advantages, including high thermal efficiency, compact size, lightweight design, affordability, and ease of maintenance. These features have made plate heat exchangers a preferred choice in various industries.

In plate heat exchangers, hot and cold fluids flow through a series of plates. The unique geometry of these plates induces turbulence, enhancing heat transfer capabilities. This increased turbulence leads to a significant improvement in thermal efficiency. As a result, plate heat exchangers have become extensively employed in crucial sectors such as oil, gas, petrochemicals, and steelmaking.

The utilization of laser cutting technology in the fabrication of plate heat exchangers enables precise and intricate designs. Laser cutting ensures the accurate production of complex plate geometries, facilitates the creation of efficient flow channels, and maximizes heat transfer efficiency. Moreover, laser cutting allows for the customization of plate heat exchangers to meet specific operational requirements and optimize performance.

The broad range of applications for plate heat exchangers highlights their significance in various industries, where efficient heat transfer is vital. The integration of laser cutting in the manufacturing process of plate heat exchangers plays a crucial role in enhancing their performance and reliability. This underscores the importance of laser cutting technology in the heat transfer domain.

Plate heat exchangers often consist of plates with a low thickness, typically less than 1.2 mm, to enhance heat transfer efficiency [15]. Consequently, laser cutting is the preferred method for cutting these plates. Alternative cutting techniques, such as air cutting, tend to cause significant distortion in thin plates. Wirecutting, on the other hand, is a slow process that is unsuitable for efficient production lines. Therefore, the implementation of laser cutting technology in plate heat exchanger manufacturing enables the production of plates with exceptional precision, quality, and speed.

Figure 2.3 visually depicts the stages of the laser cutting process for heat exchanger plates, including the fluid inlet and outlet ports, which are critical components of the plate. Laser cutting ensures the precise and accurate cutting of these sensitive areas as well as the surrounding plate material. It is worth noting that, thanks to the efficiency of laser cutting, each plate can be cut in less than 1 minute, significantly reducing production time.

The utilization of laser cutting in the fabrication of plate heat exchanger plates enables the achievement of high accuracy and quality while maintaining the required thinness for optimal heat transfer performance. This advanced cutting method plays a vital role in meeting the demanding standards of plate heat exchanger production. The result is an efficient and reliable heat transfer system that caters to the specific needs of industries such as oil, gas, petrochemicals, and steelmaking.

2.3 TYPES OF LASER CUTTING PROCESSES

In addition to laser cutting, various cutting methods are utilized in the industry. It is essential to understand these alternative methods in order to appreciate the unique benefits and advantages of laser cutting. The following section provides a brief overview of some commonly employed cutting techniques, followed by a comprehensive analysis of the distinctive characteristics and advantages offered by laser cutting.

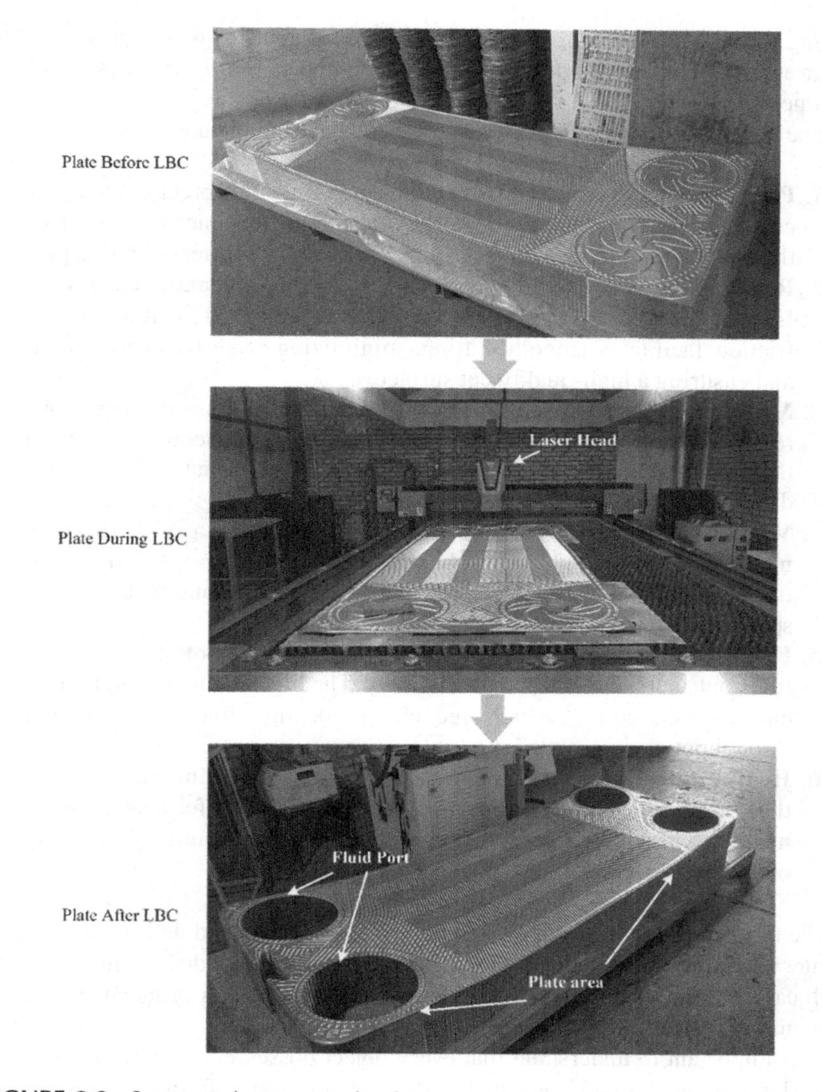

FIGURE 2.3 Laser cutting process in plate heat exchangers before, during, and after laser cutting.

2.3.1 Ultrasonic Cutting

Another notable industrial cutting method is ultrasonic cutting. This technique employs an ultrasonic cutter blade that oscillates at frequencies ranging from 20,000 to 40,000 times per second (20–40 kHz). The oscillation of the blade generates a powerful acoustic wave, enabling the cutting of various materials. The primary objective of utilizing oscillation is to minimize friction during the cutting process, allowing for easier and more efficient cutting with reduced power requirements. Ultrasonic cutting is commonly employed for thin plates with low thickness.

The key features and advantages of ultrasonic cutting include:

1. **Precision and clean cuts**: Ultrasonic cutting ensures precise and clean cuts, particularly in delicate materials. The high-frequency oscillation of the blade allows for accurate cutting along intricate patterns and shapes.
2. **Reduced friction**: The oscillation of the blade significantly reduces the friction between the cutting tool and the workpiece. This reduction in friction facilitates smooth cutting, minimizing material deformation, and ensuring a high-quality cut surface.
3. **Minimized force requirements**: The use of ultrasonic vibrations reduces the force required for cutting. This results in reduced strain on the equipment and enables the cutting of materials that are otherwise difficult to process with conventional methods.
4. **Versatility**: Ultrasonic cutting can be employed for a wide range of materials, including plastics, rubber, textiles, and composites. Its versatility makes it suitable for various industries, such as automotive, aerospace, packaging, and textiles.
5. **Enhanced productivity**: The precision and efficiency of ultrasonic cutting contribute to enhanced productivity. The high-frequency oscillation enables faster cutting speeds, reducing production time, and increasing throughput.
6. **Heat minimization**: Ultrasonic cutting produces minimal heat during the cutting process. This is particularly advantageous for heat-sensitive materials, as it helps to prevent thermal damage and ensures the integrity of the cut material.

While ultrasonic cutting is highly effective for thin plates and delicate materials, it may have limitations when it comes to cutting thicker or denser materials. In such cases, alternative cutting methods, such as laser cutting or mechanical cutting, may be more suitable.

It is important to understand that every object possesses a natural frequency, and when excited with a frequency that matches its natural frequency, resonance takes place. This principle is utilized in ultrasonic cutting machines for cutting various parts. In this method, an AC alternating current is supplied to a piezoelectric transducer. The transducer, in turn, causes the blade of the device to oscillate at its natural frequency, resulting in resonance. To ensure precise control over the cutting process, a feedback control system is employed to regulate the size and frequency of the blade. This control mechanism ensures that the desired acoustic wave is generated for effectively cutting the workpiece.

Figure 2.4 illustrates both the schematic and real implementation of this cutting method. Part a depicts the conceptual diagram, highlighting the interaction

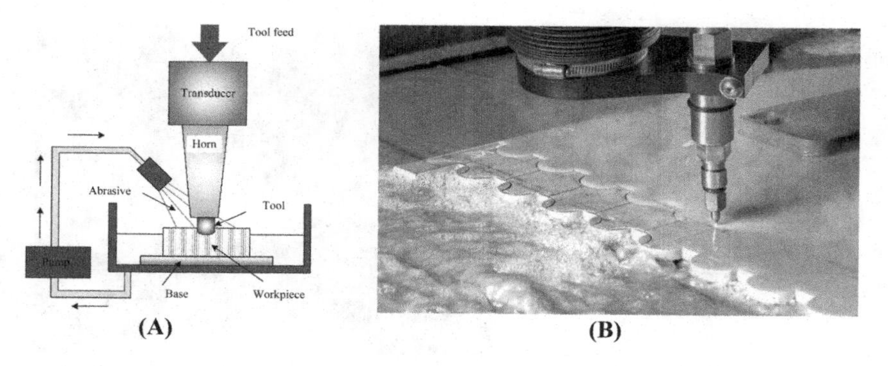

FIGURE 2.4 Ultrasonic cutting process: (a) schematic, (b) actual (real).

between the piezoelectric transducer, the blade, and the workpiece. Part b of Figure 2.4 showcases the practical application of the method, showcasing the components involved in the ultrasonic cutting process.

By harnessing resonance and employing a feedback control system, ultrasonic cutting machines enable efficient and precise cutting of various materials. The utilization of the natural frequency of the blade through the piezoelectric transducer ensures optimal cutting performance, contributing to the overall effectiveness of the ultrasonic cutting process.

2.3.2 Oxyflame Cutting

An alternative cutting method utilized in the industry is commonly referred to as "air cutting" or "oxygen flame" cutting, and it is also known by other names such as "air gas" and "oxyacetylene" cutting. This technique involves heating the metal workpiece to its combustion temperature. For steel parts, this temperature is approximately 960°C (specific values may vary depending on the alloy composition). At this elevated temperature, the steel loses its resistance to oxygen while remaining in a solid state.

During the air-cutting process, a nozzle directs a stream of pure oxygen into the heated area. The pressurized oxygen stream induces the melting and expulsion of the heated steel, effectively cutting the material through an annealing process. The oxygen plays a dual role by facilitating the exothermic reaction and assisting in the removal of excess materials.

Figure 2.5 provides both a schematic and an actual view of the air-cutting process. Part a of the figure illustrates the conceptual diagram, showcasing the interaction between the heated workpiece, oxygen stream, and cutting nozzle. Part b of Figure 2.5 offers a real-world perspective, visualizing the practical implementation of the air cutting technique.

It is worth noting that air cutting, or oxygen flame cutting, is considered one of the more economical cutting methods available. The combination of its

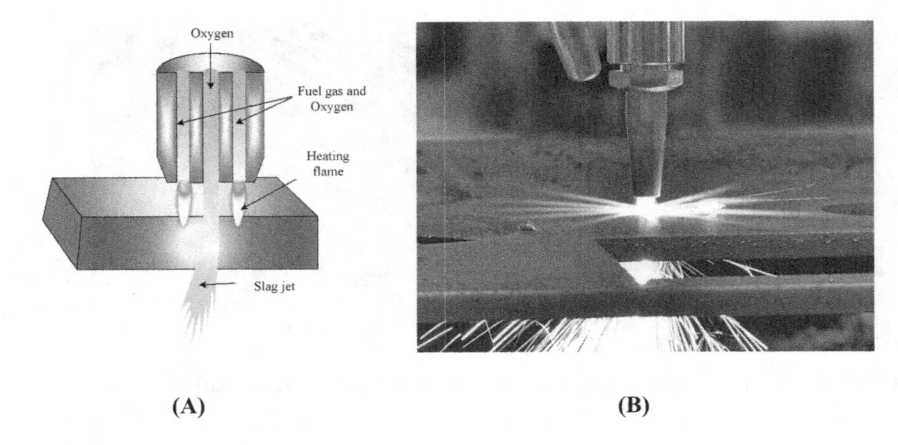

FIGURE 2.5 Air cutting process: (a) schematic view and (b) actual view.

cost-effectiveness and the ability to cut through various metal workpieces makes it a popular choice in many industrial applications.

2.3.3 CNC Milling

Cutting methods can be broadly categorized into two groups: those that involve filing and those that do not. Sampling methods, on the other hand, can be classified based on the type of movement employed: rotational or linear. Milling machines are a prime example of chipping machines that utilize rotational motion. In the case of milling, the cutting process is achieved through the movement of the blade. In today's era of technological advancement, the integration of computers and computational devices has permeated the realm of milling. This development has significantly enhanced the speed and precision of the machining process.

Figure 2.6 provides both a schematic and an actual representation of a computer numerical control (CNC) milling machine. Part a of the figure illustrates the conceptual diagram, showcasing the key components and the working principle of the machine. Part b of Figure 2.6 offers a real-world perspective, providing an actual view of the CNC milling machine in operation.

The incorporation of computer control in milling machines, known as CNC milling, has revolutionized the manufacturing industry. CNC milling machines offer greater automation, higher precision, and increased productivity. The utilization of computerized control enables the execution of complex cutting operations with remarkable accuracy and efficiency.

2.3.4 Wire Electrical Discharge Machine or Wire EDM

The wire electric discharge cutting method, commonly known as wirecut, has been in use for over half a century. This technique relies on a thermoelectric

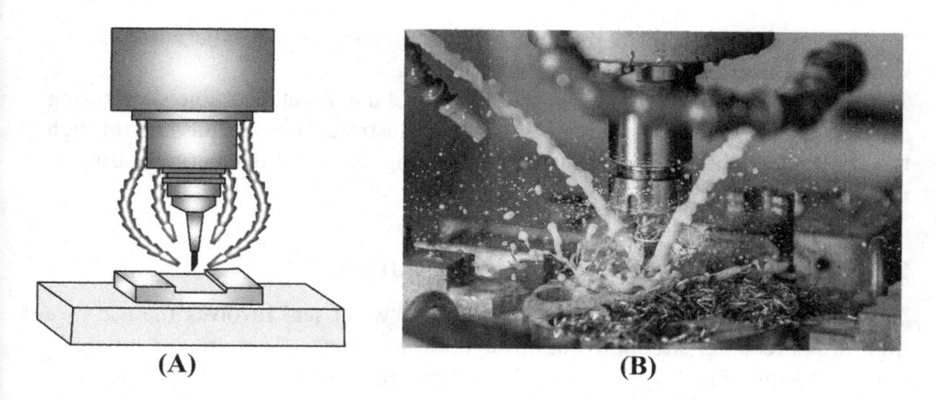

FIGURE 2.6 CNC milling cutting machine during the cutting process: (a) schematic and (b) real.

energy source to facilitate the removal of chips. It operates by generating controlled alternating sparks between the electrodes, namely the wire and the workpiece, thereby inducing the cutting process in the workpiece.

In wirecutting, a thin electrode wire, typically around 25 μm in diameter, is unwound from a spool and guided through the workpiece. The wire is continually fed through the workpiece using a mechanism, as depicted in part a of Figure 2.7. To facilitate the cutting process, a small gap is maintained between the wire and the workpiece, which is filled with a dielectric liquid during the cutting operation. When an appropriate voltage is applied, electrical discharge takes place between the wire, the workpiece, and the dielectric. This discharge creates sparks, causing localized evaporation of the workpiece material. The dielectric liquid serves the purpose of flushing away the debris, thus facilitating the chipping process.

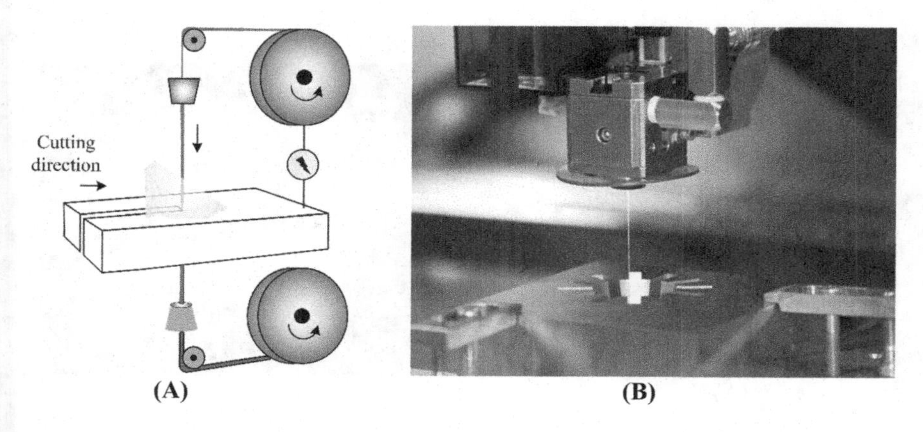

FIGURE 2.7 Wirecut cutting process: (a) schematic view and (b) actual view.

Notably, in this method, there is no physical contact between the wire and the workpiece, ensuring precise and delicate cutting.

Part b of Figure 2.7 provides an actual view of a wirecut in action, showcasing the process of cutting as it occurs. The image illustrates the wire passing through the workpiece, with the sparks generated during electrical discharge leading to material removal.

2.3.5 Abrasive Water Jet or Water Jet Cutting

The cutting method commonly referred to as "water jet" involves the use of a high-pressure water stream to cut through the workpiece. This process utilizes a powerful pump that directs water at high pressure through a small-diameter nozzle. In addition, an abrasive material is introduced into the water stream, typically at a velocity of approximately 900 m/s. Silicon carbide or silicon is commonly used as the abrasive material in water jet machines.

The addition of an abrasive material enhances the cutting capability of the water jet, enabling it to cut through hard materials effectively. The water and abrasive mixture is expelled from the nozzle at a pressure ranging from 4,000 to 6,200 bar, which ensures efficient cutting of the workpiece. Figure 2.8 provides both a schematic and a real-world view of this cutting method. Part a of Figure 2.8 illustrates the setup, with water being combined with the abrasive material and directed through the nozzle under high pressure. Part b of Figure 2.8 presents an actual view of the cutting process in action, showcasing how the high-pressure water stream, combined with the abrasive material, cuts through the workpiece.

By utilizing water under high pressure and incorporating an abrasive material, the water jet cutting method offers versatility and precision for cutting a wide range of materials.

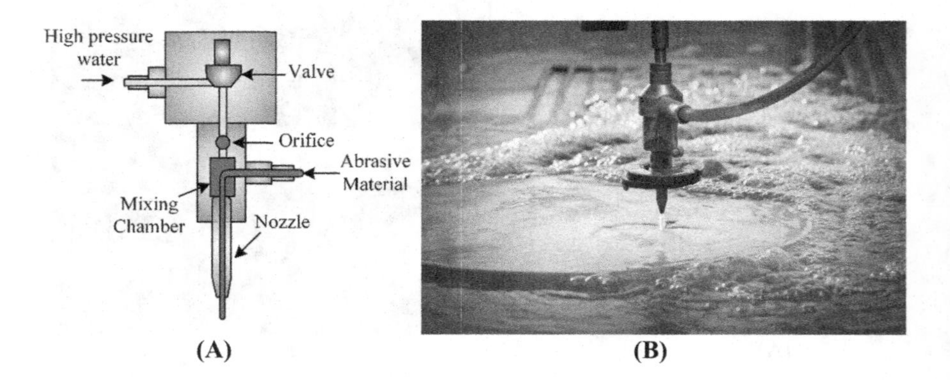

FIGURE 2.8 Water jet cutting process: (a) schematic view and (b) actual view.

2.3.6 Plasma Cutting

Plasma cutting is a cutting method that involves the utilization of an electric arc passed through a gas to cut through the workpiece. This process begins by blowing a high-speed gas, such as air, nitrogen, argon, or oxygen, from the nozzle. Simultaneously, an electric arc is generated between the gas at the nozzle head and the surface of the workpiece, causing the gas to transform into plasma. It is important to note that plasma represents the fourth state of matter, characterized by a quasi-neutral gas composed of ionized particles. In a plasma, a mixture of positive, negative, and neutrally charged particles coexist. It is worth mentioning that in plasma, the number of free electrons is nearly equal to the number of positive ions. Some scientists theorize that, following the Big Bang, matter initially existed in a plasma state. For a visual depiction of the plasma state, refer to Figure 2.9.

The schematic representation of the plasma cutting method can be observed in part a of Figure 2.10. An electric arc is established between the negative pole (torch tip) and the positive pole (workpiece). This polarity is achieved by connecting the torch and the workpiece to a high-voltage device. The potential difference between the negative and positive charges ionizes the air between the two poles,

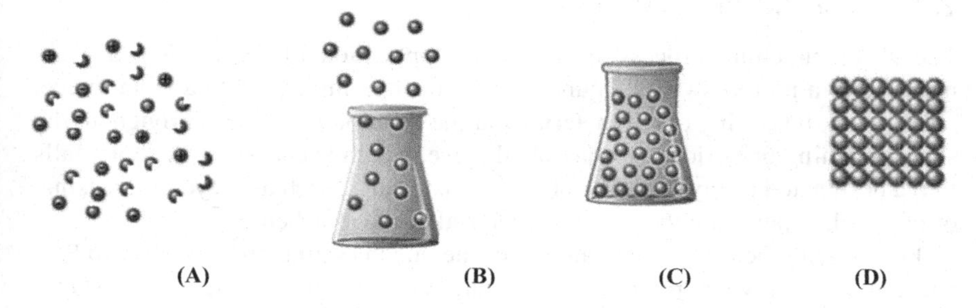

FIGURE 2.9 Four states of matter: (a) solid, (b) liquid, (c) gas, and (d) plasma.

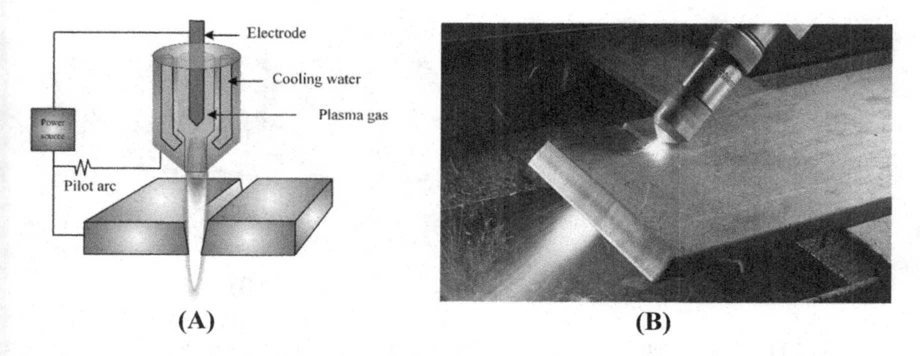

FIGURE 2.10 Plasma cutting process: (a) schematic view and (b) actual view.

resulting in the formation of plasma. The temperature of the plasma is sufficiently high to melt the metal of the workpiece. Plasma cutting offers significant advantages in terms of its cutting speed and ability to cut through various conductive materials. The high temperature of the plasma enables efficient and precise cutting, making it a widely used technique in industrial applications.

In plasma cutting, the process involves the use of a shielding gas that is directed through additional channels in the nozzle. This shielding gas is strategically positioned along the cutting path to spray the molten material away, facilitating the cutting process. It is important to note that when dealing with thin sheets, a single gas flow (comprising a combination of gas and fuel) is sufficient to effectively expel the molten material. However, when cutting thicker sheets, two separate gas flows are required. In such cases, the shielding gas options can include nitrogen, nitrogen/hydrogen, oxygen, argon, or argon/hydrogen. These gas combinations serve to optimize the cutting process for different materials and thicknesses.

For a visual representation of the plasma cutting process, refer to part b of Figure 2.10, which provides an actual view of the cutting operation. This image depicts the interaction of the plasma and the workpiece, showcasing the effectiveness of the plasma cutting method in action.

2.3.7 Punch-Cutting Process

The sheet punching process involves the application of force using a press machine to a tool known as a punch. This punch is inserted into a metal piece, resulting in the cutting of the material and the creation of a hole. Throughout the sheet punching operation, a waste metal piece, which is cut out of the hole, falls into a designated cavity. The fate of these waste parts, such as recycling, reusing, or disposal, depends on the type and material of the punched component.

For a comprehensive understanding of the punch cutting process, refer to Part a of Figure 2.11, which provides a schematic view illustrating the various stages and components involved in the sheet punching operation. Part b of Figure 2.11

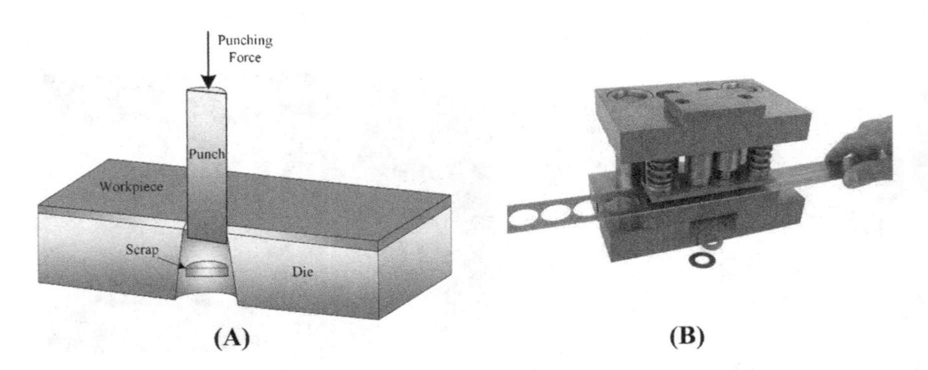

FIGURE 2.11 Punch cutting process: (a) schematic view and (b) actual view.

offers an actual view, allowing observers to visualize the punch in action, cutting through the metal piece and generating the desired hole. This visual representation enhances the comprehension of the punch cutting process and its practical implementation.

2.3.8 Nibbling Process Cutting

Nibbling is a unique cutting process characterized by the creation of a closed contour. It involves the utilization of a punching machine equipped with a small circular tool. When the punching operation is performed continuously and the resulting holes overlap, it leads to the formation of a cut known as nibbling. During nibbling, the workpiece is subjected to a series of continuous blows at a rate typically ranging from 300 to 900 blows per minute. The nibbling process employs a punch and a small die, as depicted in part a of Figure 2.12. In essence, nibbling can be regarded as a specialized variant of the punching technique, enabling precise and controlled cuts along a closed contour.

For a visual representation, refer to part b of Figure 2.12, which showcases an example of a nibbling machine. This machine demonstrates its capability to cut a 2.5-mm-thick sheet with the power of two horsepower. By observing this illustration, readers can gain a better understanding of the nibbling process and appreciate its applications in various industries.

2.3.9 Sawing and Cutting Process

The saw machine is widely recognized as one of the simplest and most efficient methods for creating grooves and cuts in workpieces within the industrial realm. This process involves a saw blade with teeth shaped as small wedges that are

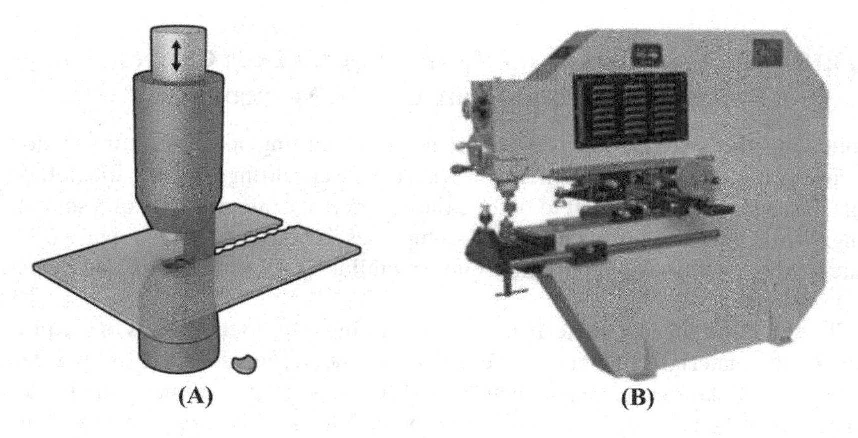

FIGURE 2.12 Nibbling cutting process: (a) schematic view and (b) real view of the device.

FIGURE 2.13 Saw cutting process: (a) actual view and (b) schematic view.

arranged in a sequential manner, allowing them to remove material from the workpiece and facilitate cutting. Typically, the workpiece is securely held in place during the cutting operation. A schematic representation of this process can be observed in part a of Figure 2.13.

It is worth noting that there are approximately 26 different types of sawing machines available. In the context of Figure 2.13, the depicted machine corresponds to the "printing saw" variant. Part a of Figure 2.13 provides a visual illustration of this specific type. For a more realistic portrayal, refer to part b of Figure 2.13, which showcases the actual view of a specialized device belonging to this category. By exploring the schematic and real views presented in Figure 2.13, readers can gain valuable insights into the operation of saw machines and appreciate their significance in various industrial applications. These machines offer a convenient and effective solution for achieving precise cuts and grooves in workpieces.

2.3.10 The Advantages of the Properties of the Laser Cutting Process Compared to Other Cutting Methods

Following the introduction of several industrial cutting methods, this section focuses on comparing these methods with the laser cutting process. To initiate this discussion, it is essential to outline the distinctive characteristics of laser cutting. While the laser cutting process encompasses various properties, Figure 2.14. categorizes them into three main groups: metallurgical, geometrical, and defect characteristics.

The metallurgical characteristics of laser cutting refer to the effects of the process on the material being cut. These include considerations such as heat-affected zone (HAZ), hardness, microstructure, and thermal stress. Geometrical characteristics, on the other hand, pertain to the precision and accuracy achieved in the cut, including factors like kerf width, taper, and cut edge quality. Lastly, the defect

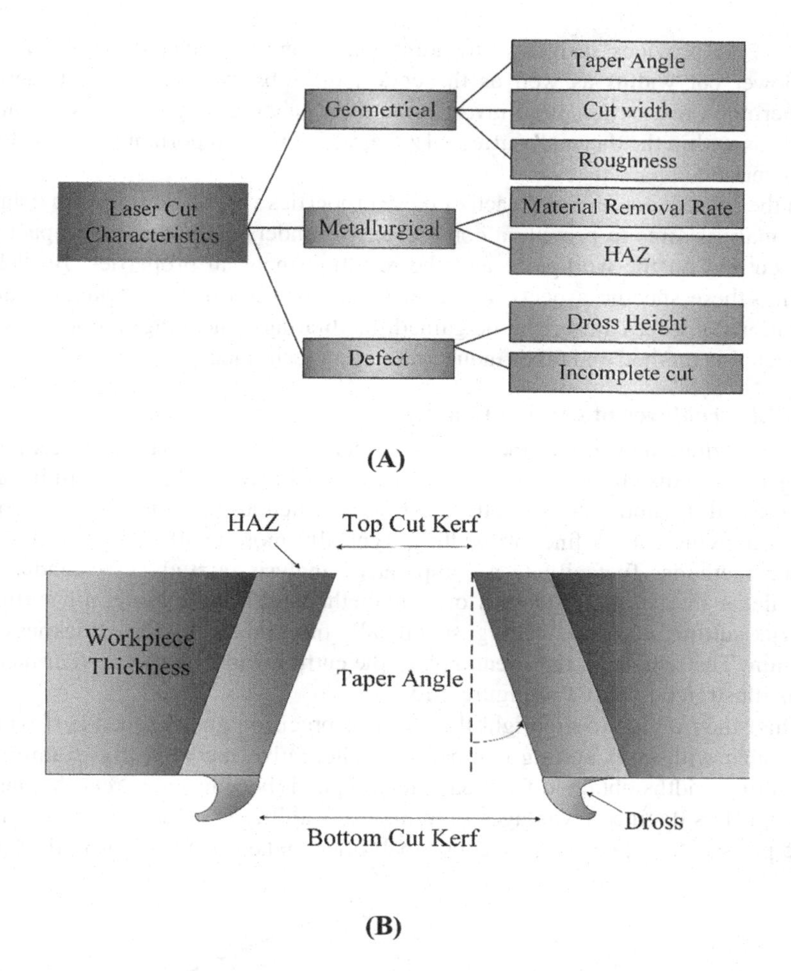

FIGURE 2.14 Properties of laser cutting process: (a) classification of its types and (b) geometric characteristics.

characteristics encompass aspects such as cutting speed, productivity, and material utilization efficiency. By evaluating the aforementioned characteristics within the context of Figure 2.14., a comprehensive understanding of the laser cutting process can be attained. This will enable a meaningful comparison between laser cutting and other industrial cutting methods.

According to the information presented in part a of Figure 2.14, several characteristics of the laser cutting process warrant consideration, namely the taper angle, cutting width, and surface roughness. These characteristics play a crucial role in assessing the quality and precision of laser-cut materials. Part b of

Figure 2.14 provides insights into additional properties, including the upper and lower cut widths as well as the cone angle observed in laser-cut plates. Furthermore, researchers have investigated the discharge rate of microstructural materials within the thermally affected zone, which is an important metallurgical characteristic.

In the subsequent sections, each of these properties will be examined in detail. This analysis aims to provide a comprehensive understanding of the impact of laser cutting on the workpiece and the resulting material properties. By delving into these specific aspects, a deeper comprehension of the advantages and limitations of laser cutting can be gained, facilitating a meaningful comparison between this method and other industrial cutting techniques.

2.3.10.1 Features of Cutting Quality

In laser cutting, the width of the cut, often referred to as the kerf, can be remarkably narrow. This characteristic not only preserves material but also minimizes damage and disruption to the cutting surface, particularly in thin plates. In contrast, achieving such a fine cut width is typically more challenging with other cutting methods. To facilitate a comparative analysis, a study was conducted to evaluate the cutting outcomes of four methods: plasma cutting, air cutting, waterjet cutting, and laser cutting, specifically on mild steel with a thickness of 12.7 mm. The schematic representation of the cutting widths for these four methods is illustrated in part a of Figure 2.15.

This study aimed to highlight the superior precision and reduced kerf width achievable with laser cutting compared to alternative methods. By examining the cutting widths obtained from each technique, it becomes apparent that laser cutting offers distinct advantages in terms of its ability to achieve narrower and more precise cuts. Such fine-cutting capabilities make laser cutting particularly

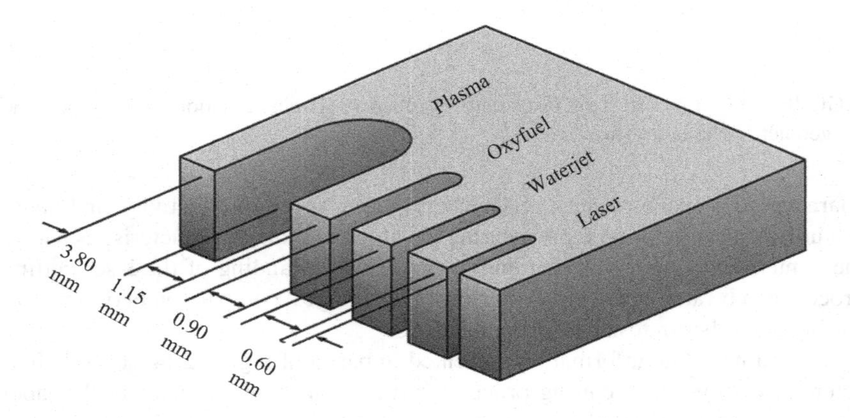

FIGURE 2.15 Comparison of laser cutting processes with other cutting methods from the point of view of cutting width.

valuable in applications that require intricate designs, minimal material waste, and high-quality finished products.

Another notable aspect to consider is the cutting quality in terms of the laser cutting edges compared to waterjet, plasma, and punch processes. Part b of Figure 2.15 illustrates the cutting edge characteristics resulting from the laser process, which is distinguished by its smoothness, straightness, and square shape. In contrast, the edges produced by waterjet and plasma methods often exhibit non-smooth and circular geometries. A study conducted on a workpiece made of S355J2+N steel with a thickness of 10 mm examined the vertical tolerance of various cutting processes, as depicted in the study [16]. The investigation revealed that the laser cutting process achieved the lowest tolerance, approximately 0.12 mm.

To further compare the accuracy of the punch cutting method with laser cutting, the study [1] presents a comparison of the resulting cutting edges from both techniques. These comparisons demonstrate the superior precision and tight tolerance achievable with laser cutting.

In addition to the cutting quality, another distinguishing factor between laser cutting and other methods is the thermal effective zone (HAZ). In a separate study, the HAZ was examined for laser and plasma cutting processes on workpieces with a thickness of 5 mm using S960MC, S700MC, and S500MC steels [17]. The laser cutting conditions included a laser power of 4,000 W, a cutting speed of 6.5 m/min, a nozzle diameter of 0.8 mm, a distance of 0.7 mm between the nozzle and workpiece, and a focal distance of 1.8 mm. The plasma cutting process was performed at 100 A, a speed of 2,159 mm/min, and a torch-to-workpiece distance of 4 mm. The results indicated that while the thermal effective zones varied for the different steels, the HAZ resulting from the laser cutting process was consistently smaller than that of the plasma process (approximately 1.3 for the laser cutting process of S500 steel compared to the plasma process).

2.3.10.2 Process Characteristics

The laser cutting process stands out as one of the fastest and most precise cutting methods available. Higher cutting speeds contribute to reduced production time, leading to increased overall production rates. Figure 2.16 illustrates that both laser and plasma cutting processes are considered high-speed cutting methods. It is important to note that the provided diagram represents the maximum power of 2,000 W, and as the power increases, laser cutting can be performed at even higher speeds. Furthermore, laser cutting exhibits versatility, as it is not limited to specific workpiece materials. Whether the material is brittle, conductive, electrically insulated, or falls on the spectrum of soft to hard, the laser cutting device can effectively cut through it.

According to Table 2.2, laser cutting demonstrates good and acceptable accuracy across various materials such as aluminum, steel, superalloy, titanium,

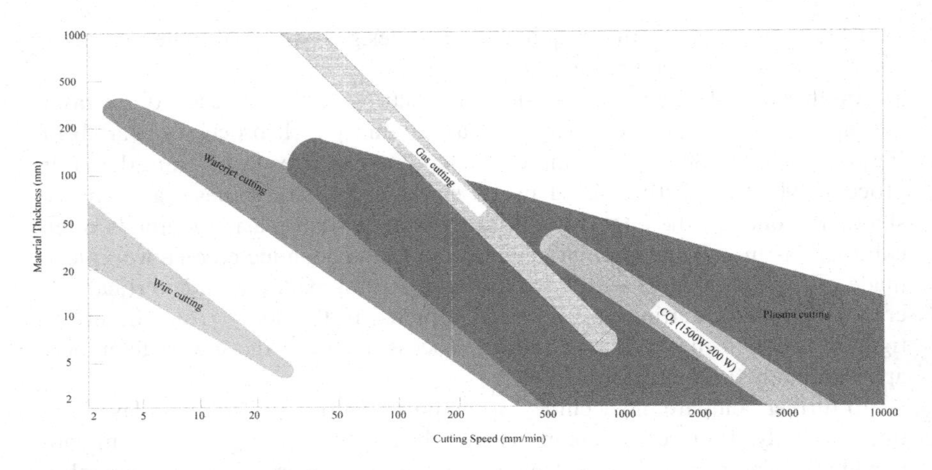

FIGURE 2.16 Comparison of laser cutting processes with other cutting methods from the point of view of thickness according to cutting speed.

TABLE 2.2
Comparison of the Use of Each Cutting Process According to Materials

Cutting Material	Al	Steel	Super alloy	Ti	Plastic	Ceramic	Glass
Ultrasonic	C	B	C	B	B	A	A
Wirecut	B	B	A	B	B	A	A
LBC	B	B	B	B	B	A	A
Plasma	A	A	A	B	C	D	D

Note: "A" indicates good, "B" means acceptable, "C" represents weak and "D" is not acceptable.

plastic, ceramic, and glass, surpassing other cutting methods in terms of precision. Only materials that possess high light reflectivity, such as aluminum, copper, and gold, present some challenges in laser cutting. However, these challenges can be easily addressed by implementing measures such as controlling the cutting beam and appropriately adjusting the lenses [18].

The laser cutting process offers several advantages, starting with the fact that it is a non-contact method. As a result, the tools used in laser cutting are not subjected to physical damage, leading to reduced equipment repair costs. However, it is important to note that maintaining clean lenses is crucial for optimal performance. Therefore, the only service cost associated with laser cutting pertains to lens cleaning.

One notable aspect of laser cutting is its unrestricted cutting direction. It allows cutting to be performed in any desired direction, providing greater flexibility compared to other cutting methods. Laser cutting settings primarily rely on software, minimizing the need for complex hardware adjustments. This feature enhances the adaptability of laser cutting to the specific requirements of the workpiece and cutting conditions, making it a highly versatile option [2].

REFERENCES

[1] C. L. Caristan, *Laser Cutting Guide for Manufacturing*. Society of Manufacturing Engineers, 2004.

[2] W. M. Steen and J. Mazumder, *Laser Material Processing*. Springer, 2010.

[3] A. Kar, D. L. Carroll, W. P. Latham, and J. A. Rothenflue, "Cutting performance of a chemical oxygen-iodine laser on aerospace and industrial materials," *Journal of Laser Applications*, vol. 11, no. 3, pp. 119–127, 1999.

[4] M. Dell'Erba, L. Galantucci, and S. Miglietta, "An experimental study on laser drilling and cutting of composite materials for the aerospace industry using excimer and CO2 sources," *Composites Manufacturing*, vol. 3, no. 1, pp. 14–19, 1992.

[5] R. Brown and R. Frye, "High-brightness laser cutting & drilling of aerospace materials," *International Congress on Applications of Lasers & Electro-Optics*, vol. 1996, no. 1, pp. C78–C85, 1996.

[6] A. Felske and F. Lunzmann, "A robot laser as a flexible tool for cutting openings in the car-body on the production line," *Laser Processing: Fundamentals, Applications, and Systems Engineering*, vol. 668, pp. 282–287, 1986.

[7] C. Anghel, K. Gupta, and T. Jen, "Analysis and optimization of surface quality of stainless steel miniature gears manufactured by CO_2 laser cutting," *Optik*, vol. 203, p. 164049, 2020.

[8] G. Wall, H. Podbielska, and M. Wawrzynska, *Functionalised Cardiovascular Stents*. Woodhead Publishing, 2018.

[9] N. Muhammad, M. M. Al Bakri Abdullah, M. S. Saleh, and L. Li, "Laser cutting of coronary stents: Progress and development in laser based stent cutting technology," *Key Engineering Materials*, vol. 660, pp. 345–350, 2015.

[10] C. Momma, U. Knop, and S. Nolte, "Laser cutting of slotted tube coronary stents-state-of-the-art and future developments," *Progress in Biomedical Research*, vol. 4, no. 1, pp. 39–44, 1999.

[11] H. Meng, J. Liao, Y. Zhou, and Q. Zhang, "Laser micro-processing of cardiovascular stent with fiber laser cutting system," *Optics & Laser Technology*, vol. 41, no. 3, pp. 300–302, 2009.

[12] X. Lin, P. Wang, Y. Zhang, Y. Ning, and H. Zhu, "Theoretical and experimental aspects of laser cutting using direct diode laser source based on multi-wavelength multiplexing," *Optics & Laser Technology*, vol. 114, pp. 66–71, 2019.

[13] S. Marimuthu, J. Dunleavey, Y. Liu, M. Antar, and B. Smith, "Laser cutting of aluminium-alumina metal matrix composite," *Optics & Laser Technology*, vol. 117, pp. 251–259, 2019.

[14] M. Sharifi and M. Akbari, "Experimental investigation of the effect of process parameters on cutting region temperature and cutting edge quality in laser cutting of AL6061T6 alloy," *Optik*, vol. 184, pp. 457–463, 2019.

[15] B. Sundén and R. M. Manglik, *Plate Heat Exchangers: Design, Applications and Performance*. Wit Press, 2007.

[16] M. S. Węglowski and T. Pfeifer, "Influence of cutting technology on properties of the cut edges," *Advances in Manufacturing Science and Technology*, vol. 38, no. 2, pp. 63–73, 2014.
[17] C. Jiménez-Peña, C. Goulas, J. Preußner, and D. Debruyne, "Failure mechanisms of mechanically and thermally produced holes in high-strength low-alloy steel plates subjected to fatigue loading," *Metals*, vol. 10, no. 3, p. 318, 2020.
[18] A. Akkurt, "The effect of cutting process on surface microstructure and hardness of pure and Al 6061 aluminium alloy," *Engineering Science and Technology, an International Journal*, vol. 18, no. 3, pp. 303–308, 2015.

3 Laser Beam Cutting Process

Machine, Mechanism, and Processing Parameters

3.1 LASER CUTTING MACHINE

The development of laser cutting machines represents a significant integration of principles from light physics, laser engineering, and numerical control engineering. The origins of these machines can be traced back to the 1970s, when a combination of laser technology and punch cutting techniques was introduced. This combination marked a pivotal moment in the evolution of laser cutting machines, setting the stage for their subsequent advancements.

A typical laser cutting machine comprises several key components, each playing a crucial role in the cutting process. These components include the resonator or laser source, beam transmission system, laser motion control system, cooling system, and gas transmission system [1]. Figure 3.1 provides both a schematic and real views of a standard laser cutting machine, offering a visual representation of its components and structure. In the following sections, we will explore each of these components in detail, unveiling their functions and significance in

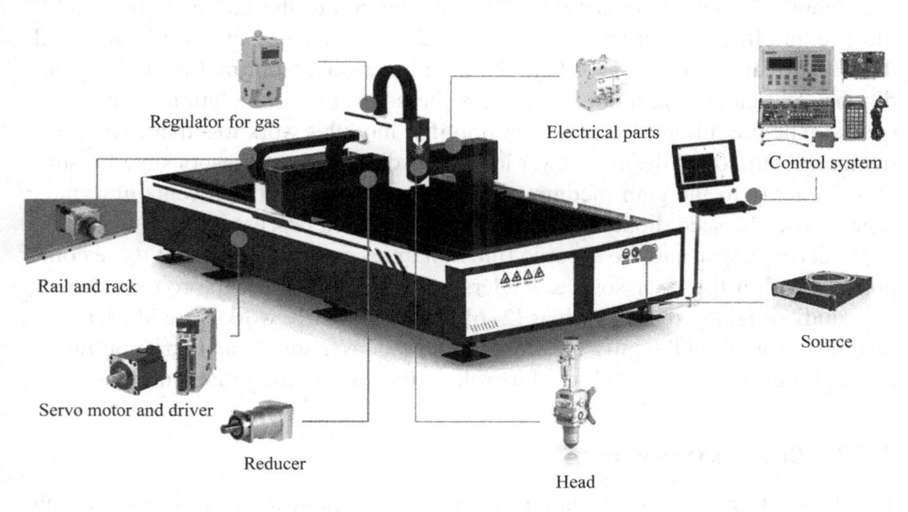

FIGURE 3.1 Laser cutting machine and its parts.

DOI: 10.1201/9781003492191-3

facilitating the laser cutting process. This comprehensive understanding is essential for grasping the intricacies of laser cutting technology and its applications in various industries.

3.1.1 Resonator or Laser Source

In the context of a laser cutting machine, the laser resonator, often referred to as the "source," assumes a pivotal role as it is responsible for generating the laser beam. Essentially, it serves as the core component of the entire laser cutting system. The laser resonator is recognized as the most financially significant part of the machine, given its central role in the cutting process. Its construction comprises various components, each contributing to its operation and overall efficiency. Parts a and b of Figure 3.2 provide a detailed depiction of these components, offering a comprehensive visual reference for a better understanding of the intricate workings of the laser resonator.

Understanding the inner workings and components of the laser resonator is crucial for comprehending the principles that drive laser cutting technology. As we delve further into the details of each constituent part, we gain insights into how the laser resonator generates and maintains the laser beam, setting the stage for its effective utilization in the cutting process. This knowledge lays the foundation for grasping the broader functionalities of a laser cutting machine and its significance in various industrial applications.

The laser source, often referred to as the "resonator," stands as the central and financially significant component in the laser cutting machine, responsible for the generation and amplification of the laser beam. Its critical role in the laser cutting process makes understanding its internal components crucial for comprehending the overall functionality of the machine.

Among the key components of the laser source are the fan, electronic module, pre-amplifier, power amplifier module, diode pump, output isolator, and seed diode. The fan is essential for heat dissipation, ensuring optimal operating conditions. The electronic module manages the control and regulation of the laser source. The pre-amplifier and power amplifier modules work together to amplify the laser beam to the desired power level. The diode pump is responsible for supplying energy to the gain medium, while the output isolator prevents unwanted reflections. The seed diode initiates the laser emission process.

To delve deeper into the specific functionalities and interactions of these components within the laser source, readers are encouraged to explore comprehensive studies and references such as [2–6]. These scholarly works provide detailed insights, explaining the intricate workings of the laser source and contributing to a comprehensive understanding of its role in the laser cutting machine.

3.1.2 Beam Transfer System

The beam transmission system is a crucial component of the laser cutting machine, serving as a conduit for transferring the laser beam from the resonator to the workpiece. This system is instrumental in preserving the fundamental

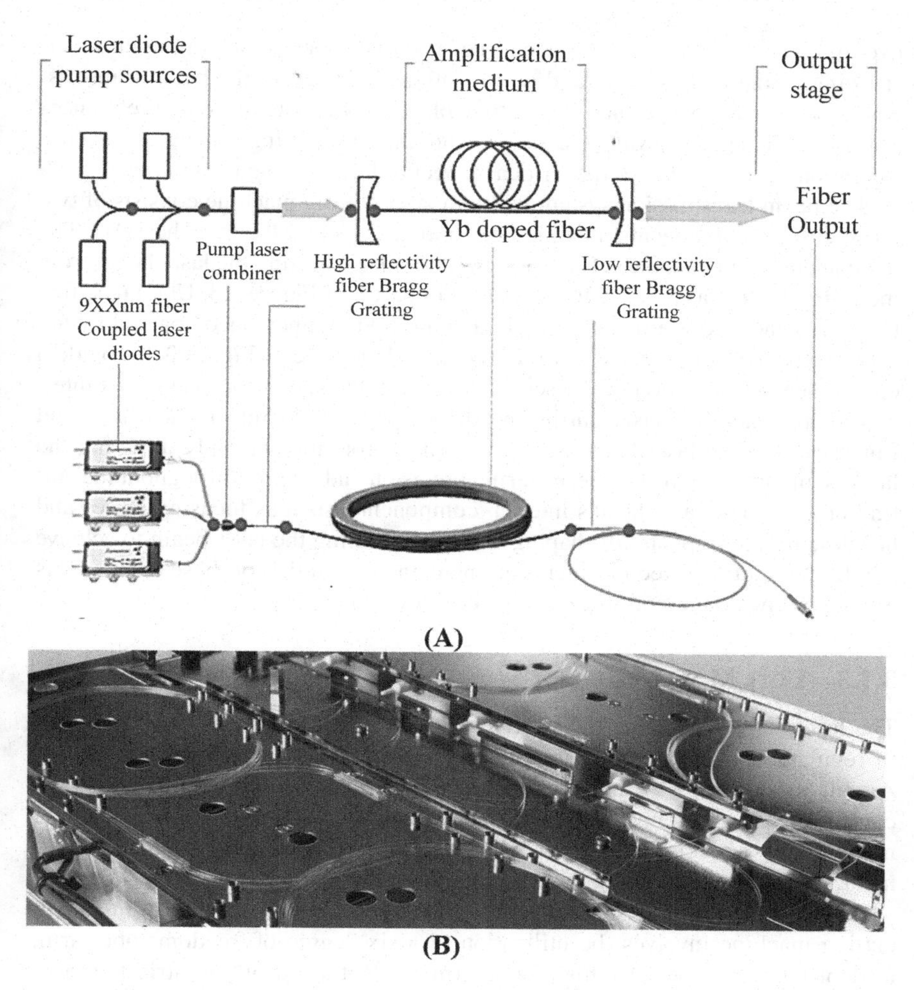

FIGURE 3.2 Laser cutting machine source: (a) block diagram and configuration and (b) real parts.

characteristics of the laser beam, including power, diameter, and concentration, during the transmission process. The significance of maintaining these attributes cannot be overstated, as any deviation can directly impact the efficiency and precision of the cutting process. To achieve optimal cutting performance, it is essential to minimize potential losses and ensure that the laser beam remains in its desired state throughout the transmission.

To comprehend the intricate workings of the beam transmission system, it is essential to consider its key elements, which may include mirrors, lenses, fiber optics, and other optical components. These components collectively contribute to directing and focusing the laser beam onto the workpiece with precision. The arrangement, alignment, and quality of these elements are paramount to the overall effectiveness of the laser cutting machine. For a more in-depth understanding of the complexities and optimizations involved in the beam transmission system, readers are encouraged to explore relevant literature and references [24–28], which provide valuable insights into the engineering principles and considerations associated with this critical aspect of laser cutting technology.

The beam transmission system within a laser cutting machine consists of two indispensable components: flexible fiber optic cables and the laser head [1]. The flexible optical fiber cable acts as the conduit for transmitting the laser beam from the source to the laser head, as showcased in part b of Figure 3.3. Upon reaching the laser head, the beam is emitted in a controlled manner, accompanied by the carrier gas, onto the surface of the workpiece. Parts a–c of Figure 3.3 provide a comprehensive depiction of the schematic representation, actual view, and internal components of a laser cutting head, offering insight into its structure and functionality. The laser head assumes a critical role in accurately directing the laser beam onto the workpiece, ensuring precision and control throughout the cutting process. The laser head's internal components, such as focusing lenses and nozzles, are instrumental in shaping and concentrating the laser beam to achieve the desired cutting effect. Proper alignment and calibration of these components are imperative to maintaining cutting accuracy and quality.

3.1.3 Laser Motion Control System

The laser cutting machine incorporates diverse control methods, each designed to regulate critical process parameters, ensuring precise and efficient cutting. Among these control methods, computer numerical control, programmable logic control (PLC), and robotic control stand out as prominent approaches [1]. These methods play a pivotal role in managing key parameters such as laser beam alignment, machine power, cutting speed, and the distance to the workpiece.

One notable example showcasing the integration of robotic control in a laser cutting machine involves the utilization of a six-degree-of-freedom robot arm, as depicted in Figure 3.4. This robotic arm is adept at executing intricate movements, and its six degrees of freedom provide flexibility and versatility in directing the laser beam. Servomotors, responsible for supplying the necessary energy for arm movements, ensure precision and controlled motion, contributing to the overall accuracy of the cutting process.

For linear movements, alternative systems like shoulder-rotator or wing-screw mechanisms equipped with servomotors can serve as viable substitutes for robotic arms, depending on the specific requirements of the application. This adaptability highlights the versatility of control methods within the laser cutting machine, offering tailored solutions to accommodate diverse cutting needs. For a detailed exploration of these control methods and their applications, interested readers can refer to relevant studies and resources in the field [1].

In summary, the integration of various control methods enhances the adaptability and precision of laser cutting machines, making them versatile tools for a wide range of applications. Whether employing computer numerical control,

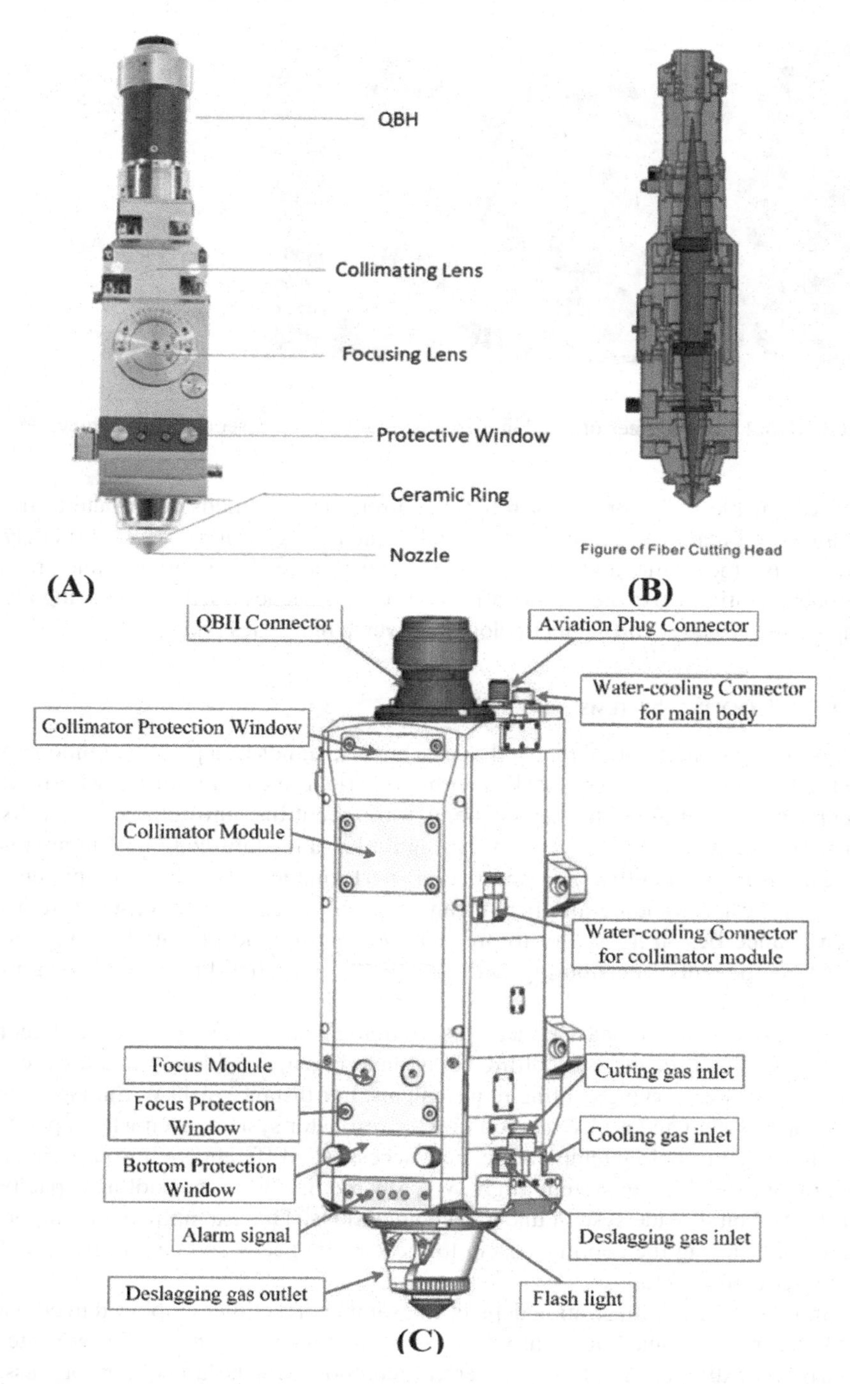

FIGURE 3.3 Head of the laser cutting machine: (a) real, (b) schematics, and (c) parts.

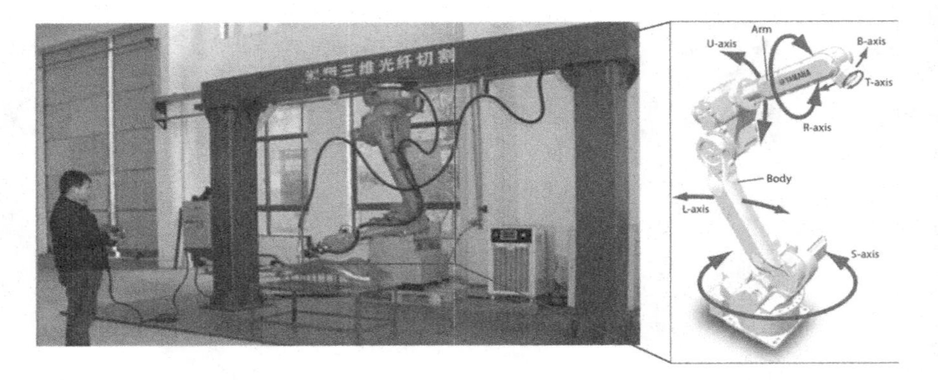

FIGURE 3.4 Six degrees of freedom robot arms are used in laser cutting machines.

programmable logic control, or robotic control, these methods contribute to the seamless regulation of parameters critical to the laser cutting process, ultimately influencing the quality and efficiency of the cuts produced. Researchers and practitioners continue to explore innovative control approaches, further advancing the capabilities of laser cutting technology in diverse industries [1].

3.1.4 Cooling Systems

The cooling system, often referred to as a chiller, stands as a pivotal component in the laser cutting machine, tasked with the critical role of maintaining optimal operating temperatures for key elements throughout the cutting process [1]. Its primary function revolves around the regulation and stabilization of temperatures, ensuring the efficiency and overall performance of crucial components such as the laser source and beam transmission system. The cooling system's significance lies in its ability to prevent overheating and potential damage to these components, contributing to the longevity and reliability of the laser cutting machine.

It is essential to emphasize that the selection of a cooling system is contingent on the specific type of laser cutting being undertaken, considering that different lasers may have varying cooling requirements. For instance, a 1 kW Endiag laser resonator system and a 1 kW carbon dioxide resonator system, even when operating at the same cooling temperature, may necessitate different cooling capacities. In practice, the Endiag system might typically require twice the cooling capacity of the carbon dioxide system under such conditions. This exemplifies the importance of tailoring the cooling system to the specific needs and characteristics of the laser cutting setup.

In Figure 3.5, an illustration depicts an example of a chiller employed to cool a 4 kW laser cutting machine. This visual representation highlights the chiller's integral role in maintaining optimal temperature conditions for the laser cutting process. By efficiently dissipating excess heat generated during laser operation, the cooling

FIGURE 3.5 The chiller used in cooling the 4kW laser cutting machine.

system ensures that the laser cutting machine operates within its designed temperature range, thereby optimizing performance and contributing to the precision and quality of the cuts produced.

For a more in-depth understanding of the intricacies of cooling systems and their applications in laser cutting machines, readers are encouraged to refer to comprehensive studies and resources within the field [1]. These resources provide valuable insights into the considerations and factors involved in selecting and implementing cooling systems tailored to diverse laser cutting requirements.

3.1.5 Gas Transmission System

The gas transfer system in a laser cutting machine serves the crucial function of expelling the molten material from the cutting area [1]. This process involves two key steps: the laser beam melts the targeted area, and then gas is employed to propel the molten material outward. The choice of gas utilized in this system can vary depending on the specific laser cutting method employed.

To facilitate the transfer of the gas, a flexible or rigid tube is employed to deliver the gas to the nozzle. Rigid pipes are typically constructed from materials such as copper or stainless steel, providing durability and efficient gas flow. On the other hand, flexible pipes are made from materials like polyethylene, nylon, or Teflon, offering greater flexibility and ease of maneuverability in the gas transfer process [1].

These materials are selected for their compatibility with the gases used and their ability to withstand the operational conditions of the laser cutting machine.

3.2 TYPES OF LASER CUTTING

From a comprehensive perspective, cutting lasers can be broadly categorized into three main types: gas lasers, crystal lasers, and fiber lasers. Each of these categories possesses distinct characteristics and operational principles, which will be elucidated in further detail in the subsequent sections.

Gas lasers encompass laser systems that employ various types of gas as the lasing medium. These gases, such as carbon dioxide (CO_2), nitrogen (N_2), or helium-neon (He-Ne), generate the laser beam through the excitation of gas molecules. Gas lasers are well-known for their high power output and efficiency in industrial cutting applications.

Crystal lasers, on the other hand, utilize solid-state crystals, such as neodymium-doped yttrium aluminum garnet (Nd:YAG), as the lasing medium. These crystals are typically optically pumped to create the necessary population inversion for laser emission. Crystal lasers are valued for their ability to produce high-energy laser beams with exceptional beam quality and stability.

Fiber lasers represent a more recent advancement in laser technology. They employ an optical fiber doped with rare-earth elements, such as erbium (Er) or ytterbium (Yb), as the gain medium. Fiber lasers offer notable advantages, including high efficiency, compact size, and excellent beam quality. They have gained significant popularity in various cutting-edge applications.

In subsequent sections, the specific characteristics, operational principles, and applications of gas, crystal, and fiber lasers will be explored in greater detail, providing a comprehensive understanding of these cutting-edge laser technologies.

3.2.1 GAS LASERS

When the first cutting laser was developed in 1964, its power was initially insufficient for cutting metals effectively. However, significant advancements have been made since then, enabling gas lasers, particularly carbon dioxide lasers, to be employed for metal cutting applications. Although gas lasers are commonly utilized for non-metal cutting, they are now capable of cutting metals as well. These lasers typically operate at a wavelength of 10.6 μm and find widespread use in both medical and industrial fields.

In the context of gas lasers, nitrogen is occasionally employed as a cutting gas, specifically for cutting metals like steel and aluminum. It is crucial to ensure the purity of the nitrogen used in the process. Impure nitrogen may induce metal oxidation during cutting, thereby affecting the quality and integrity of the cut. Hence, stringent measures are necessary to maintain the purity of the nitrogen gas utilized.

For further insights into the intricate details of laser cutting technology, referring to studies [7–9] would be highly beneficial. These studies delve into various aspects of laser cutting, offering comprehensive information and facilitating a deeper understanding of this advanced manufacturing technique.

3.2.2 Crystal Lasers

Laser crystals, also known as solid-state laser materials, are a type of laser medium that utilizes solid crystalline materials to generate and amplify laser light. These crystals are carefully chosen based on their optical and electronic properties to facilitate the production of coherent and intense laser beams.

Laser crystals can be made from various materials, including crystalline substances such as ruby (aluminum oxide doped with chromium), neodymium-doped yttrium aluminum garnet (Nd:YAG), neodymium-doped yttrium orthovanadate (Nd:YVO_4), and erbium-doped yttrium aluminum garnet (Er:YAG). These crystals are selected based on their ability to absorb pump energy and efficiently emit laser light at specific wavelengths.

The choice of laser crystal depends on the desired laser output characteristics, such as the operating wavelength, output power, pulse duration, and beam quality. Each crystal type has unique properties that make it suitable for specific applications, ranging from industrial cutting and welding to medical procedures, scientific research, and telecommunications.

Laser crystals undergo a process called "doping," where impurities or dopant atoms are intentionally added to the crystal lattice to alter its optical properties. The dopant atoms interact with the crystal lattice and allow for stimulated emission, leading to laser amplification and light emission.

By carefully selecting and optimizing the crystal material, laser designers can achieve precise control over laser output characteristics, making laser crystals a crucial component in various laser systems across multiple industries.

One of the problems with these lasers is the expensiveness of their diode pumps. Also, these lasers have a short lifespan compared to other lasers. Their useful life is approximately 8,000–15,000 hours. This is the reason that, after this period of time, the diode pump must be replaced. For a better understanding of crystal cutting lasers, you can refer to studies [10,11].

3.2.3 Fiber Lasers

A fiber laser is a type of laser that uses an optical fiber as the laser medium or gain medium. It operates on the principle of stimulated emission of radiation, similar to other types of lasers. However, unlike traditional gas or crystal lasers, which use a solid or gas medium, a fiber laser utilizes a doped optical fiber to generate the laser beam.

The core of a fiber laser is a specially designed optical fiber that is doped with rare-earth elements such as erbium, ytterbium, or neodymium. These dopants are chosen for their ability to amplify light at specific wavelengths. The fiber is typically made of silica glass, which provides a high-quality optical waveguide.

The functioning of a fiber laser involves the excitation of the dopant atoms within the fiber core. This excitation can be achieved through various means, such as injecting pump light into the fiber or using diode lasers as the pump source. The excited dopant atoms then emit photons through stimulated emission, which leads to the amplification of light within the fiber.

The light generated within the fiber is reflected back and forth between the ends of the fiber, undergoing further amplification through the interaction with the dopant atoms. This process is known as optical feedback or optical resonance. The laser beam is extracted from the fiber through one or both ends, usually using specialized optics.

Fiber lasers offer several advantages over other types of lasers. They are known for their high efficiency, excellent beam quality, compact size, and robustness. Fiber lasers are widely used in various applications, including material processing (such as cutting, welding, and marking), telecommunications, scientific research, medical devices, and defense systems. Due to their compactness and versatility, fiber lasers have gained significant popularity and have become a preferred choice in many industries where precision and reliability are essential [35–37].

3.2.4 Conclusion of Laser Cutting Machines

Based on a research report [3], it was found that in the American market in 2016, fiber lasers accounted for 61.03% of cutting lasers, followed by carbon dioxide lasers at 14.70%, solid-state (crystal) lasers at 13.29%, and gas lasers at 10.98%. The popularity of fiber lasers can be attributed to their high efficiency compared to other types of lasers.

Fiber lasers have a distinct advantage in terms of power output. In fact, fiber lasers can produce light with nearly twice the power of carbon dioxide lasers. This higher power output makes fiber lasers well-suited for a wide range of cutting applications. Moreover, the method of delivering the laser light to the cutting site is simpler and more straightforward with fiber lasers. Unlike other types of lasers, fiber lasers do not require expensive optical lenses or intricate moving parts, resulting in a more cost-effective and efficient system.

The efficiency and simplicity of fiber lasers have contributed to their widespread adoption and have led to a decline in the market growth of carbon dioxide and solid-state lasers in recent years. This trend highlights the superior performance and advantages offered by fiber lasers, making them the preferred choice for many industrial applications.

It is worth noting that the aforementioned information is based on the research report [3], which provides valuable insights into the market dynamics and preferences of laser technologies in the United States.

3.3 PHYSICAL MECHANISMS OF LASER CUTTING

According to various academic studies, laser technology offers a range of cutting methods that utilize different physical mechanisms during the cutting process.

These methods can be categorized into seven main approaches: (1) evaporative cutting, (2) fusion, (3) reactive fusion, (4) controlled fracture, (5) scribing or scraping, (6) cold cutting and (7) laser-assisted oxygen cutting. These cutting methods demonstrate the versatility of lasers and their ability to adapt to various material properties and cutting requirements. The choice of method depends on factors such as the material being cut, desired precision, cutting speed, and the specific application. Details of these methods will be provided below:

3.3.1 Evaporative Cutting

Evaporative laser cutting is a cutting method that relies on the heating effect of the laser beam to create a hole in the cutting area. This process involves several stages that facilitate the penetration and cutting of the workpiece. Initially, the laser beam heats the surface of the cutting area, raising its temperature to the boiling point. This localized heating creates a hole in the material, which aids in the laser beam's penetration. As the hole deepens, more material is melted. The molten material on the cavity wall is subsequently evaporated by the intense heat of the laser beam. This evaporation enlarges the cavity, leading to the final separation and cutting of the workpiece. It is crucial to note that in this method, the objective is not to keep the workpiece in a molten state but rather to vaporize it. Evaporative laser cutting is commonly employed for non-metallic materials such as wood, carbon, and thermoset plastics. Figure 3.6 provides a schematic view of this cutting method, illustrating the sequential stages involved. By understanding the principles and mechanisms behind evaporative laser cutting, researchers and practitioners can optimize this method for various non-metallic materials, ensuring precise and efficient cutting processes [12].

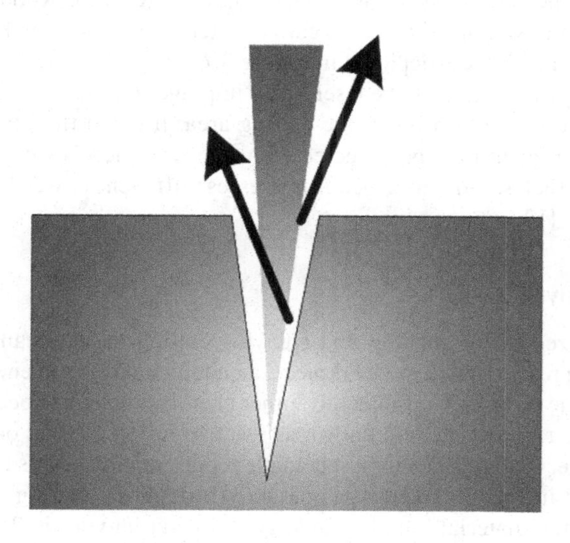

FIGURE 3.6 Schematic view of laser cutting by vaporization method.

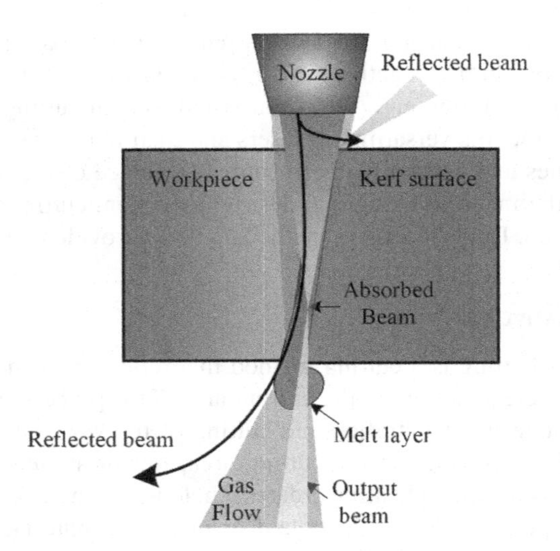

FIGURE 3.7 Schematic view of laser cutting by melting and spraying.

3.3.2 Fusion Cutting

In laser fusion cutting, also referred to as "melting and sputtering," the energy delivered by the laser creates a hole in the material, initiating the cutting process. Simultaneously, gas is directed towards the cutting area, effectively expelling the molten material. Unlike the previous method discussed, laser fusion cutting does not require the workpiece to reach its vaporization point or exceed its melting point. Consequently, less energy is needed for the cutting process. A comparative analysis has indicated that laser fusion cutting requires approximately half the energy of evaporative laser cutting. To gain a better understanding of this method, refer to the schematic view depicted in Figure 3.7.

As illustrated in Figure 3.7, the laser beam impinges on the workpiece, causing significant material removal from the cutting area. It is worth noting that some of the incoming beam may be reflected from the workpiece's surface or molten material; nevertheless, the laser beam possesses sufficient power to accomplish the cutting task [12].

3.3.3 Reactive Fusion

If the gas utilized in the melting and spraying cutting method can engage in a heat-generating reaction with the workpiece, an additional thermal energy source is introduced to the process. Consequently, other chemical reactions occur at the cutting site, giving rise to a method known as reactive fusion cutting, or "melt, burn, and spray." Refer to Figure 3.8 for a schematic representation of this method [12].

As depicted in Figure 3.8, the gas passing through the cutting area not only expels the molten material but also undergoes a reaction with it. Typically, oxygen or a combination of gases containing oxygen is employed as the reactive gas

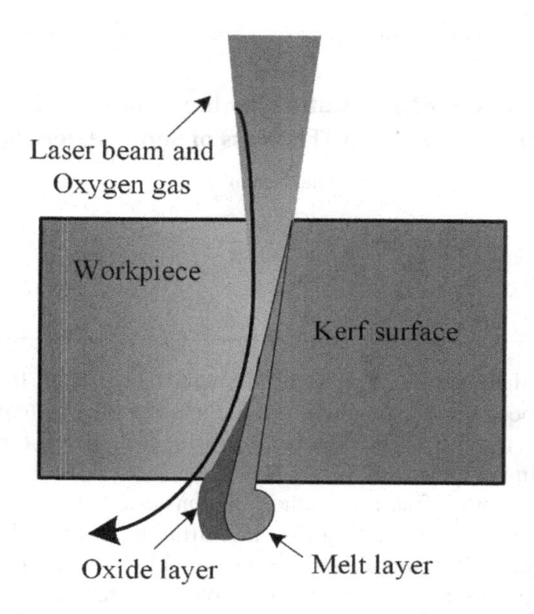

FIGURE 3.8 Schematic view of laser cutting by the reactive fusion method.

in this method. The combustion process typically initiates at the surface when the temperature surpasses the ignition point. Consequently, oxidation occurs and extends towards the interior of the cutting site, driving the molten material downward. The molten material acts as a protective coating, mitigating the chemical reaction with oxygen. This controlled reaction serves to accelerate the cutting speed, with the cutting speed in this method being approximately twice that of the melting and spraying methods [12].

3.3.4 Controlled Fracture

Brittle and fragile materials, which are susceptible to thermal failure, can be effectively and precisely fractured by concentrating heat on them. This concept serves as an introduction to the controlled fracture method of laser cutting using thermal stress. In the controlled fracture method, also known as "thermal stress cracking," the laser beam heats a small area on the surface of the workpiece. The application of heat causes the workpiece to expand, leading to the generation of tensile stress. If a crack exists within the desired boundary, the concentration of stress will result in the propagation of the crack in the direction of the heat movement (laser beam). It is important to note that the crack growth rate is relatively rapid, approximately 1 m/s. Ultimately, the crack will reach the lower edge of the workpiece, initiating the cutting process.

The controlled fracture method exhibits impressive speed, quality, and accuracy when applied to materials such as glass. However, it presents challenges when cutting closed profiles. In this method, the speed of crack propagation

TABLE 3.1
The Power Required for Laser Cutting by the Controlled Breaking Method Is Suitable for the Material and Thickness of Some Materials

Material	Thickness (mm)	Input Power (W)
Quartz	0.8	3
Soda glass	1	10
Sapphire	1.2	12
Al_2O_3 99%	7	7

exceeds that of the laser beam, making it somewhat difficult to control cuts in closed profiles positioned at angles, like rectangular shapes (e.g., car mirrors). The cut advances faster than the laser beam, posing a challenge for precise control in complex cutting geometries.

While the controlled fracture method demonstrates remarkable capabilities, it is essential to consider its limitations in cutting closed profiles. The original research paper provides comprehensive insights into this laser cutting method, including its applications, limitations, and potential areas of improvement [12]. In the controlled fracture process, the surface of the workpiece should not melt, as this may damage the cutting edge. Therefore, according to this point, it is expected that the required power of this method is low. For a better understanding, in Table 3.1, the required power of some materials according to their material and thickness is listed.

3.3.5 Scribbling or Scraping

The laser scribing or scratching process involves creating a groove or cut in the form of a line with varying levels of permeability. The permeability can be high enough that the cutting site can be easily broken by hand or mechanically. This laser cutting method finds application in materials where the thermal area is minimal and the chips generated are very small. It is particularly useful for cutting silicon chips and alumina layers. Essentially, the cutting and scribing method shares similarities with the evaporation method but requires less energy for operation [12].

It is worth emphasizing that the small thermal area and the generation of tiny chips make the laser scribing or scratching process suitable for delicate materials such as silicon chips and alumina layers. By referring to the cited research paper, one can gain a deeper understanding of the intricacies and applications of this laser cutting method.

3.3.6 Cold Cutting

The cold-cut laser cutting process is a relatively new method that originated with the development of high-power excimer lasers. These lasers emit a beam in the ultraviolet range, characterized by extremely short pulse durations, typically in the picosecond range. The energy of the ultraviolet photons generated by these lasers is approximately 4.9 eV, which happens to be comparable to the binding energy of many organic substances. When these organic materials are exposed to the stimulated ultraviolet photons, the bonds within them may break.

In the cold cutting process, the cutting occurs when these material bonds fail in the direction of the desired cut. Unlike other laser cutting methods, this technique operates without generating excessive heat, thereby avoiding melting. Instead, the workpiece is sheared without undergoing a phase change. Due to its ability to provide precise cutting without thermal damage, the cold cutting method finds applications primarily in medical and surgical equipment manufacturing [12].

It is worth noting that the cold-cut laser cutting process represents a significant advancement in laser technology, offering an alternative approach for cutting organic materials with minimal thermal impact. For further insights into this method and its specific applications, the mentioned research paper can be consulted.

3.3.7 Laser-Assisted Oxygen Cutting

The laser cutting process can be further enhanced by introducing oxygen into the cutting path, enabling the cutting of thick materials with lower laser power. This method is commonly referred to as "oxygen cutting with laser" or "laser cutting with stabilizer burning. "By employing this technique, even a relatively low-power laser, similar to a match, can effectively cut through thick materials.

For instance, a 1 kW laser is capable of cutting steel with a thickness of 50 mm at a speed of 200 mm/min. Notably, laboratory experiments have demonstrated successful cutting of thicknesses up to 80 mm using this method. Figure 3.9 provides a schematic representation of the process [12]. It is important to highlight that oxygen-assisted laser cutting represents an efficient approach for cutting thick materials, allowing for enhanced cutting capabilities with lower laser power.

3.3.8 Comparison of Physical Mechanisms of the Laser Cutting Process

In the preceding section, we discussed different methods and physical mechanisms employed in laser cutting. To provide further insights into the practical application of each method, Table 3.2 offers a comparative analysis, considering the relative energy requirements and the applicable workpiece types.

From the data presented in Table 3.2, it is evident that the scrubbing and controlled failure mechanisms exhibit the lowest relative energy requirements. Conversely, the cold cutting mechanism demands the highest relative energy. When it comes to cutting metal parts, fusion mechanisms such as reactive fusion and oxygen cutting with lasers are commonly employed. The compiled information in Table 3.2 facilitates a comprehensive understanding of the various

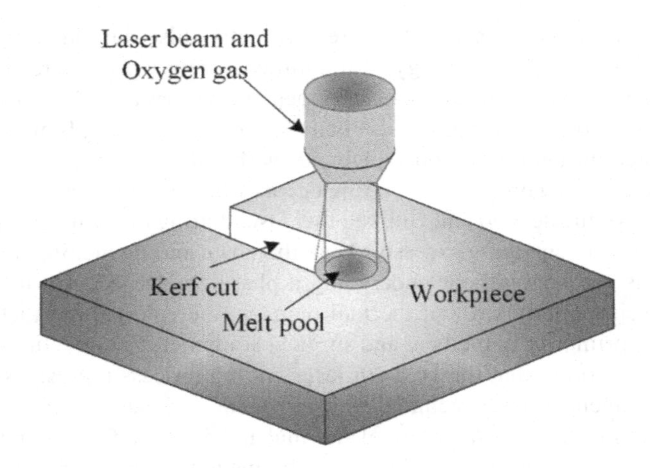

FIGURE 3.9 Schematic view of laser cutting using the oxygen cutting method with laser.

TABLE 3.2
Comparison of Different Types of Laser Cutting Mechanisms from the Point of View of Required Energy and Workpiece Material

Method	Relative Energy	Workpiece Material
Evaporation	40	Wood, carbon, thermoset plastic
Fusion	20	Metals, thermoplastics and some glasses and ceramics
Reactive fusion	10	Metals
Controlled fracture	1	Glass, sapphire
Scribing	1	Human hair
Cold cutting	100	Organic materials
Laser-assisted oxygen cutting	5	Metals

laser cutting methods by shedding light on their relative energy demands and the recommended workpiece types.

3.4 EFFECTIVE PARAMETERS OF THE LASER CUTTING PROCESS

The fusion cutting mechanism stands out as one of the most extensively used techniques in the industry, finding applications in cutting metals, thermoplastics, glasses, and ceramics [13]. Given its significance, this section delves into the influential factors in the laser cutting process using the fusion cutting mechanism.

As depicted in part a of Figure 3.10, the physics of the laser fusion cutting process can be divided into three stages: heating, melting, and spraying

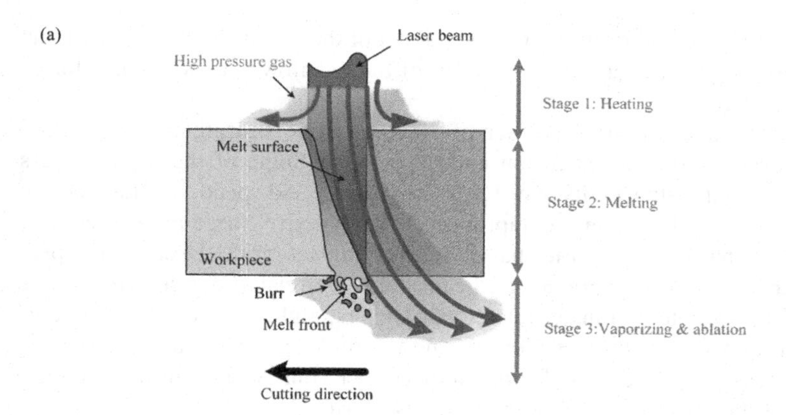

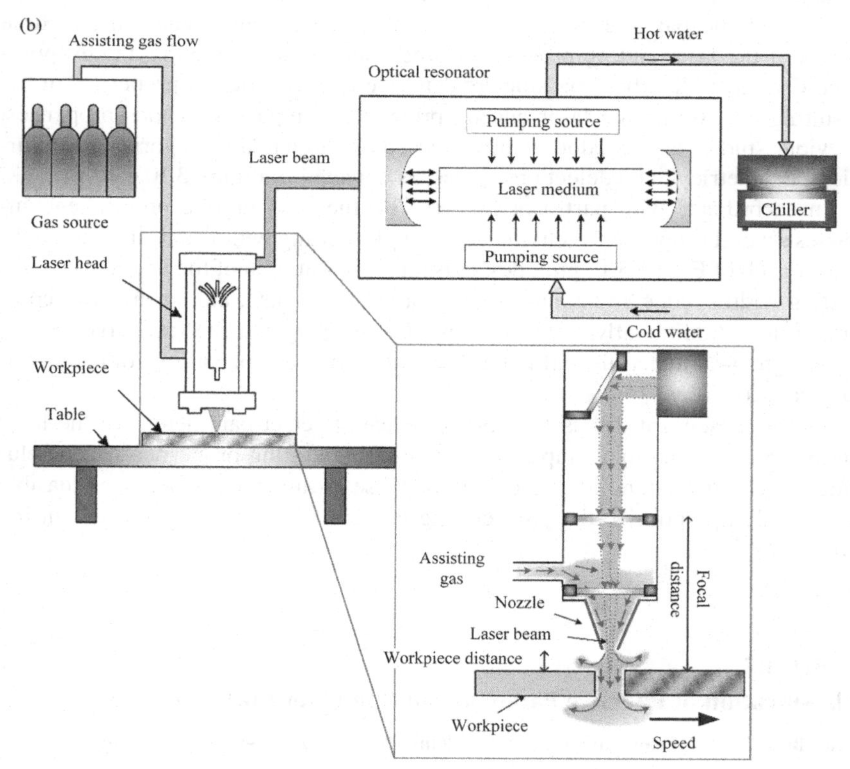

FIGURE 3.10 Laser cutting process: (a) schematic of the physics of the process and (b) schematic of the components of the laser cutting machine and effective parameters.

and evaporation. Initially, the laser beam generated in the optical resonator illuminates the workpiece, leading to its heating in the first stage. Subsequently, the workpiece undergoes melting in the second stage. High-pressure carrier gas from the gas capsule is then propelled onto the molten section, causing a portion of the melt to be expelled and resulting in the creation of the cut. It should be noted that during this stage, the melt not in the path of the gas flow evaporates and forms the cutting wall.

Various parameters influence each of the three steps (heating, melting, spraying, and evaporation) in the fusion cutting process. Some of these parameters include the focal distance, distance to the workpiece, and speed, as illustrated in part b of Figure 3.10. From a comprehensive perspective, these parameters can be classified into four categories based on their impact on the laser cutting process: laser beam-related parameters, beam guidance-related parameters, carrier gas-related parameters, transfer-related parameters, and material-related parameters, as outlined in Table 3.3. Considering the multifaceted nature of the laser cutting process and the diverse factors influencing its outcome, an in-depth understanding of these parameters and their effects is crucial.

Each of the parameters discussed earlier can have an impact on the three stages of the laser cutting process: heating, melting, and spraying and evaporation. Consequently, these parameters can give rise to different properties in the resulting cut. While the laser cutting process encompasses various properties, previous studies have predominantly focused on three main categories: metallurgical, geometrical, and defect properties, as depicted in Figure 3.10.

In accordance with part a of Figure 3.10, the laser cutting process encompasses several properties, including taper angle, cutting width, and surface roughness. Part b of Figure 3.10 provides a visual representation of the upper and lower cutting widths, cone angle, and pleat on a laser-cut plate, aiding in better comprehension. Additionally, some studies have investigated the discharge rate of microstructural materials within the thermally affected zone as one of the metallurgical properties.

In the subsequent discussion, the influence of select parameters on the laser cutting process and their impact on the resulting cutting properties will be elucidated. By understanding the effects of these parameters, it becomes feasible to control and optimize the laser cutting process to achieve the desired cutting properties.

TABLE 3.3
Classification of Effective Parameters of the Laser Cutting Process

Laser Beam	Beam Guidance	Assisting Gas	Transfer	Material
Beam's diameter	Focal distance	Chemical structure	Scanning speed	Absorption coefficient
Pulsed/continuous		Density		Density
Frequency		Viscosity		Melt temperature
Power		Pressure		Evaporation temperature
		Geometry of nozzle		Thickness
		Diameter of nozzle		
		Workpiece distance		

3.4.1 Laser Beam-Based Parameters

The study investigated the influence of laser power on crucial parameters such as the hardness and width of the thermally affected zone generated during the laser cutting process [14]. In this specific investigation, the workpiece under consideration was made of low-carbon steel, with a thickness of 1.25 mm and dimensions of 50×50 mm. Understanding the impact of laser power on these properties is essential for optimizing the laser cutting process and ensuring desired material characteristics and dimensional outcomes.

The findings from this study contribute valuable insights into the relationship between laser power and the resulting thermally affected zone, shedding light on the potential trade-offs and considerations in selecting appropriate laser power settings for specific materials and applications. For a more comprehensive grasp of the experimental setup, methodologies, and detailed results, interested readers are encouraged to refer to the complete study [14], which provides a nuanced understanding of the interplay between laser power and the thermal effects induced during laser cutting.

The cutting process was conducted using a continuous Nd:YAG laser device operating at a wavelength of 1,064 nm. The study explored a range of cutting speeds, varying from 700 to 1,500 mm/min, and employed laser powers ranging from 337 to 515 W. By examining the relationship between laser power and the resulting properties of the cut, the study aimed to gain insights into the optimal power settings for achieving desired outcomes in terms of the hardness and width of the thermally affected zone.

Based on the metallurgical images obtained at different laser powers (337, 456, and 515 W) under constant conditions of a speed of 1,500 mm/min and a pressure of 5 bar, certain observations can be made. At medium and low powers, no significant changes are observed in the microstructure. Conversely, at high power, an increase in hardness is observed due to the phase change resulting from the elevated input heat. In contrast, at low laser powers, the insufficient heat input is unable to induce a phase change, resulting in the formation of coarse grains and a decrease in hardness.

Analysis of the thermal cutting area diagram reveals that increasing laser power can lead to an expansion of the effective thermal cutting area due to the higher heating rate. The fluctuations in the diagram are attributed to variations in the amount of radiation absorbed by the workpiece surface. Consequently, if a low heat level is desired in the cutting process for the effective area, it is expected that a low laser power setting would be appropriate. These findings illustrate the correlation between laser power and the resulting metallurgical properties, such as microstructure, hardness, and thermal cutting area.

3.4.2 Beam Guidance-Based Parameters

One of the important parameters that significantly influences the laser cutting process is the focal distance. Several studies have explored the impact of this parameter on the resulting cutting properties. In a particular study, the effect of focal distance on the width of the cutting seam in CO_2 laser cutting of steel sheets with thicknesses of 1.5 and 3.5 mm was investigated [15]. The experimental setup included the use of oxygen as the carrier gas at a pressure of 100 kPa, a cutting speed of 300 mm/min, a laser power of 1,500 W, a frequency of 100 Hz, a distance to the workpiece of 1.5 mm, and a nozzle diameter of 1.5 mm.

The results of the study show that the width of the cut seam decreases initially and then increases as the focal distance is reduced. This behavior can be attributed to the variation in the laser beam diameter at the focal distance of 127 mm. When the focal distance is decreased, the laser beam diameter becomes smaller, resulting in a more focused laser beam. Consequently, the cutting seam width is reduced. However, as the focal distance continues to decrease, the laser beam moves away from its optimal focus point, leading to an increase in the width of the cutting seam. These findings highlight the influence of the focal distance on the width of the cutting seam in CO_2 laser cutting of steel sheets.

3.4.3 Assisting Gas-Based Parameters

The carrier gas parameters, encompassing factors such as nozzle diameter, geometry, chemical composition, and gas pressure, are pivotal in shaping the cutting properties of the laser fusion process [16]. The pressure of the carrier gas is instrumental in providing the requisite discharge force to the molten pool, a fundamental aspect of the laser cutting mechanism. Importantly, adjustments in gas pressure do not solely impact the discharge of molten material; they also exert an influence on various quality aspects of the cut.

A careful consideration of carrier gas parameters is essential for optimizing the laser cutting process and achieving desired outcomes in terms of cut quality, precision, and efficiency. The intricate relationship between carrier gas characteristics and the resulting cut properties underscores the need for a thorough exploration of these parameters in the context of specific materials and cutting applications. Interested readers seeking a deeper understanding of the nuances associated with carrier gas parameters and their impact on laser cutting are encouraged to refer to the comprehensive study [16], which delves into the intricacies of these factors and their implications for the laser fusion process.

The results indicate that increasing the gas pressure leads to a decrease in surface roughness and a reduction in the waste area. However, it is important to note that there is a threshold for gas pressure. Exceeding this threshold can cause the melt to freeze and result in the formation of undesirable pleats along the cut. Therefore, there is an optimal range of gas pressure that needs to be carefully controlled to achieve the desired cutting outcomes without introducing detrimental effects.

Further details and specific data regarding the influence of carrier gas parameters can be found in the referenced research paper, which provides comprehensive insights into the experimental observations and the relationship between gas pressure and cutting properties in laser fusion cutting processes.

3.4.4 The Transfer-Based Parameters

Cutting speed is a critical parameter that significantly influences the cutting properties in laser cutting processes, as underscored by various research studies. In a specific investigation, the impact of cutting speed on the quality properties of laser cutting was explored using a 1 kW fiber laser with a wavelength of 1,070 nm [17].

The findings from this study shed light on the intricate relationship between cutting speed and the resulting cut quality properties. Understanding and optimizing cutting speed is essential for achieving desired outcomes in terms of precision, efficiency, and the overall effectiveness of the laser cutting process. For a more detailed exploration of the effects of cutting speed on cutting quality properties and related insights, interested readers are encouraged to refer to the comprehensive study [17], which provides in-depth analyses and valuable information on this aspect of laser cutting.

The experiment involved a low carbon steel workpiece with thicknesses of 1 and 2 mm, a laser beam diameter of 62 μm, a power of 800 W, and variable cutting speeds ranging from 30 to 85 mm/s. The focal position was set 4 mm above the workpiece, with a distance to the workpiece of 1 mm. Oxygen was used as the carrier gas with a pressure of 1 bar.

For speeds exceeding 80 mm/s, the upper part of the cutting edge demonstrates desirable surface roughness. However, the lower part of the cutting edge remains rough. As the speed further increases, the roughness gradually extends upward from the lower part of the cutting edge. Based on this comparison, it can be concluded that a speed of 50 mm/s yields the smoothest cutting surface and ensures good cutting quality.

3.4.5 Material-Based Parameters

The laser cutting process, a versatile technology applied to various materials, demonstrates distinct behaviors influenced by specific parameters. Material type is a pivotal parameter that significantly impacts the resulting cutting properties. A comprehensive study delved into the cutting quality of four diverse materials: Ti-6Al-4V aluminum alloy, 304 steel, Inconel 625, and alumina [18]. The investigation employed nitrogen as the carrier gas with a pressure of 550 kPa, a focal distance of 127 mm, a nozzle diameter of 1.5 mm, a distance to the workpiece of 1.5 mm, a frequency of 1,500 Hz, and power ranging from 300 to 1,500 W. Cutting speeds ranged from 0.1 to 0.05 m/s.

In metals, the behavior of the cutting surface shares similarities due to comparable melting patterns. The molten metal on the surface experiences rapid solidification, induced by the cooling effect from the convection of the high-pressure carrier gas. This phenomenon is more prominent at low cutting speeds and high power densities in the laser cutting process. The significant temperature gradient generated during laser cutting leads to thermal strain, particularly in areas with brittle materials such as alumina. The result can be the formation of microcracks on the cutting surface, emphasizing the importance of avoiding high power and low cutting speeds to prevent crack formation in the cutting seam for brittle materials [18].

Understanding the intricacies of the laser cutting process is crucial for optimizing its efficiency and effectiveness. Additionally, considering specific material characteristics allows for tailoring the process parameters to achieve the desired cutting quality. For an in-depth exploration of the influence of material type on laser cutting properties and related insights, the interested reader is encouraged to refer to the comprehensive study [18], which provides a detailed analysis of the behavior of different materials during the laser cutting process.

Furthermore, beyond material type, several critical parameters influence the overall performance and quality of laser cutting. The laser cutting machine itself is a complex system composed of various components, each playing a significant role in the process. The laser source, comprising components such as the fan, electronic module, pre-amplifier, power amplifier module, diode pump, output isolator, and seed diode, serves as the core component generating the laser beam [1]. These components collectively contribute to the functionality and performance of the laser resonator, emphasizing the importance of their detailed understanding. For further insights into the components of the laser source and their roles, interested readers can refer to the extensive studies cited in references [19–23].

Additionally, the beam transmission system plays a pivotal role in facilitating the transfer of the laser beam from the resonator to the workpiece. This system ensures that the beam retains its essential characteristics, such as power, diameter, and concentration, throughout the transmission process. Optimizing the beam transmission system is crucial for maintaining the desired cutting performance and maximizing the effectiveness of the laser cutting machine. For a more detailed exploration of the beam transmission system and its significance, interested readers are encouraged to refer to relevant studies and resources in the field.

In conclusion, laser cutting technology represents a sophisticated integration of various principles, including light physics, laser engineering, and numerical control engineering. Its application spans across diverse materials, each exhibiting unique behaviors during the cutting process. Material type is a fundamental parameter influencing cutting properties, and understanding these material-specific behaviors allows for tailored optimization of the laser cutting process. Moreover, a holistic comprehension of the components and parameters within the laser cutting machine is essential for achieving precise and efficient cutting outcomes. Researchers and practitioners in the field continue to explore and advance laser cutting technology, contributing to its evolution and widespread application in various industries.

REFERENCES

[1] C. L. Caristan, *Laser Cutting Guide for Manufacturing*. Society of Manufacturing Engineers, 2004.

[2] W. M. Steen and J. Mazumder, *Laser Material Processing*. Springer, 2010.

[3] A. Kar, D. L. Carroll, W. P. Latham, and J. A. Rothenflue, "Cutting performance of a chemical oxygen-iodine laser on aerospace and industrial materials," *Journal of Laser Applications*, vol. 11, no. 3, pp. 119–127, 1999.

[4] M. Dell'Erba, L. Galantucci, and S. Miglietta, "An experimental study on laser drilling and cutting of composite materials for the aerospace industry using excimer and CO_2 sources," *Composites Manufacturing*, vol. 3, no. 1, pp. 14–19, 1992.

[5] R. Brown and R. Frye, "High-brightness laser cutting & drilling of aerospace materials," *International Congress on Applications of Lasers & Electro-Optics*, vol. 1996, no. 1, pp. C78–C85, 1996.

[6] A. Felske and F. Lunzmann, "A robot laser as a flexible tool for cutting openings in the car-body on the production line," *Laser Processing: Fundamentals, Applications, and Systems Engineering*, vol. 668, pp. 282–287, 1986.

[7] C. Anghel, K. Gupta, and T. Jen, "Analysis and optimization of surface quality of stainless steel miniature gears manufactured by CO_2 laser cutting," *Optik*, vol. 203, p. 164049, 2020.

[8] G. Wall, H. Podbielska, and M. Wawrzynska, *Functionalised Cardiovascular Stents*. Woodhead Publishing, 2018.

[9] N. Muhammad, M. M. Al Bakri Abdullah, M. S. Saleh, and L. Li, "Laser cutting of coronary stents: Progress and development in laser based stent cutting technology," *Key Engineering Materials*, vol. 660, pp. 345–350, 2015.

[10] C. Momma, U. Knop, and S. Nolte, "Laser cutting of slotted tube coronary stents-state-of-the-art and future developments," *Progress in Biomedical Research*, vol. 4, no. 1, pp. 39–44, 1999.

[11] H. Meng, J. Liao, Y. Zhou, and Q. Zhang, "Laser micro-processing of cardiovascular stent with fiber laser cutting system," *Optics & Laser Technology*, vol. 41, no. 3, pp. 300–302, 2009.

[12] X. Lin, P. Wang, Y. Zhang, Y. Ning, and H. Zhu, "Theoretical and experimental aspects of laser cutting using direct diode laser source based on multi-wavelength multiplexing," *Optics & Laser Technology*, vol. 114, pp. 66–71, 2019.

[13] S. Marimuthu, J. Dunleavey, Y. Liu, M. Antar, and B. Smith, "Laser cutting of aluminium-alumina metal matrix composite," *Optics & Laser Technology*, vol. 117, pp. 251–259, 2019.

[14] M. Sharifi and M. Akbari, "Experimental investigation of the effect of process parameters on cutting region temperature and cutting edge quality in laser cutting of AL6061T6 alloy," *Optik*, vol. 184, pp. 457–463, 2019.

[15] B. Sundén and R. M. Manglik, *Plate Heat Exchangers: Design, Applications and Performance*. Wit Press, 2007.

[16] M. S. Węglowski and T. Pfeifer, "Influence of cutting technology on properties of the cut edges," *Advances in Manufacturing Science and Technology*, vol. 2, pp. 63–73, 2014.

[17] C. Jiménez-Peña, C. Goulas, J. Preußner, and D. Debruyne, "Failure mechanisms of mechanically and thermally produced holes in high-strength low-alloy steel plates subjected to fatigue loading," *Metals*, vol. 10, no. 3, p. 318, 2020.

[18] A. Akkurt, "The effect of cutting process on surface microstructure and hardness of pure and Al 6061 aluminium alloy," *Engineering Science and Technology, an International Journal*, vol. 18, no. 3, pp. 303–308, 2015.

[19] A. E. Siegman, An introduction of Lasers. In *Lasers*. University Science Books, vol. 37, no. 208, p. 169, 1986.

[20] C. W. Billings, *Lasers: New Technology of Light. Universities Press*, 1992.

[21] M. Bertolotti, *The History of the Laser*. CRC Press, 2004.

[22] O. Svelto and D. C. Hanna, *Principles of Lasers*. Springer, 2010.

[23] F. R. Karl, *Basics of Laser Physics for Students of Science and Engineering*. Spinger, 2017.

[24] A. Alizadeh and H. Omrani, "An integrated multi response Taguchi-neural network-robust data envelopment analysis model for CO_2 laser cutting," *Measurement*, vol. 131, pp. 69–78, 2019.

[25] V. S. Balakrishnan, H. Seidlitz, M. R. Yellur, and N. Vogt, "A study on the influence of drilling and CO2 laser cutting in carbon/epoxy laminates," *Journal of Materials Research and Technology*, vol. 8, no. 1, pp. 944–949, 2019.

[26] Y. Gatut et al., "Design and fabrication of optical waveguide as directional coupler using laser cutting CO_2 on acrylic substrate," *Journal of Physics: Conference Series*, vol. 1153, no. 1, p. 012100, 2019.

[27] G. D. Gautam and D. R. Mishra, "Dimensional accuracy improvement by parametric optimization in pulsed Nd:YAG laser cutting of Kevlar-29/basalt fiber-reinforced hybrid composites," *Journal of the Brazilian Society of Mechanical Sciences and Engineering*, vol. 41, pp. 1–22, 2019.

[28] P. Joshi, A. Sharma, V. Yadava, and Y. Modi, "Nd:YAG laser cutting of ni-based superalloy thin sheet: Experimental modeling and process optimization," In: *Application of Lasers in Manufacturing: Select Papers from AIMTDR 2016*, pp. 179–207, Springer Singapore, 2019.

4 Laser Beam Welding

Phenomena and Applications

4.1 AN INTRODUCTION TO WELDING

The advent of welding has revolutionized the landscape of materials, tools, and methods employed in machinery. This pivotal technique of joining metal parts has become an indispensable process, rendering the production of modern metal parts and equipment virtually impossible without its application. The significance of welding is underscored by its projected global market value of $20.48 billion by 2025, a staggering figure that highlights its paramount importance and extensive utilization within the industry. The historical origins of welding, its inventors, and the transformative changes it has undergone over time to attain its current position serve as vital inquiries within the realm of metal production technology.

To truly understand the evolution of welding, it is crucial to delve into its historical roots. The origins of welding can be traced back to ancient times, when early civilizations utilized techniques such as forge welding and brazing to join metal components. These ancient methods relied on the application of heat and pressure to create durable bonds between metal surfaces. Over the centuries, the art of welding progressed, with notable advancements made during the Industrial Revolution. Innovators such as Sir Humphry Davy, known for his work on the arc welding process, and Nikolai Benardos, who introduced carbon arc welding, played instrumental roles in propelling welding forward.

The history of metal connections can be traced back to ancient times, when humans first conceived the idea of joining materials like stone and wood to create tools such as spears. In ancient Egypt, stone tools were connected using plaster mortar, which was employed in the construction of magnificent temples and pyramids. Such examples of connections can be found abundantly throughout ancient history, indicating that methods of joining metals were discovered much earlier than one might imagine [1].

The emergence of metal connections can be observed in significant historical periods such as the Bronze Age and the Iron Age, where attempts were made to address the challenges associated with shaping and joining metal alloys. Metalworkers had long been concerned with connecting metal plates, and one of the fundamental methods employed in metal joining was molding with sand on a metal sheet, followed by direct heating to achieve the desired shape. This method, known as casting, is one of the oldest and most basic techniques for joining metals. Another method that was used in ancient times involved placing two

DOI: 10.1201/9781003492191-4

base metals at a specific distance from each other and pouring molten metal into the gap. As the molten metal cooled and solidified, the two pieces were effectively welded together [1]. Historical evidence suggests that the ancient Egyptians even practiced iron welding. Additionally, a small golden box dating back to 2,000 years ago was discovered, where the edges were welded under pressure. Another noteworthy historical example of heat-based welding is the bronze statue of a goat found in the Qingyang Taoist Temple in Chengdu, China, which was repaired using bronze welding techniques approximately 1,000 years ago.

The Greek historian Herodotus, in the 5th century BC, mentioned Glaucus of Chios as the individual who personally invented iron welding. Further historical documents highlight the use of welding in the construction of the renowned "Delhi Metal Pillar" approximately 310 years ago in India. This monumental column, measuring 7.2 m in length with a diameter of 40.64 cm and weighing around 5,865 kg, stands as a testament to the advanced welding techniques employed at that time. Reference [2] provides a visual representation of this historic column.

The development of electric welding in the late 19th century marked a significant turning point in the history of welding. Pioneers like Auguste De Méritens and C.L. Coffin contributed to the refinement of electric arc welding techniques, allowing for more precise and efficient metal joining. Subsequent advancements in the 20th century, including the introduction of gas welding, resistance welding, and the emergence of new welding processes such as plasma arc welding and laser welding, further expanded the capabilities of this transformative technology.

Today, welding has reached an unprecedented level of sophistication and versatility. Modern welding techniques encompass a wide range of processes, including shielded metal arc welding (SMAW), gas metal arc welding, and tungsten inert gas (TIG) welding, to name a few. Furthermore, the integration of automation, robotics, and computer control systems has revolutionized welding operations, enhancing precision, productivity, and safety in industrial applications.

Although historical documentation exists, the welding techniques employed in medieval times were notably primitive and distinct from the sophisticated methods utilized in modern welding today. Conventional welding during that era involved forcefully hammering two pieces of metal together until they were joined through the heat generated at the point of contact. Following the Middle Ages, this particular form of welding disappeared and did not resurface until the 19th century. Given this historical background, it is challenging to pinpoint the exact origins and inventors of welding. Rather, the development of welding, as we know it today, is the culmination of the collective efforts of numerous researchers, scientists, and inventors over the course of many years.

Some experts trace the beginning of modern welding to the early 18th century, when Sir Humphry Davy achieved the creation of the first "electric arc" between two carbon electrodes using a battery. In 1836, Edmund Davy discovered "acetylene. "However, it wasn't until 1881 that welding was introduced, utilizing these two groundbreaking inventions. In that year, August de Méritens harnessed the heat generated by the electric arc to join metal plates. Nikolai Benardes, a Russian student of de Méritens, subsequently patented a type of electric arc welding that involved the use of a carbon rod. This marked the beginning of a rapid expansion of welding techniques. Nikolai Slavinov later proposed the utilization of metal electrodes in welding, while C.L. Coffin, an American engineer, introduced a type of arc welding wherein a coated metal electrode provided protection during the process [1].

Based on this concise historical overview, it becomes evident that welding has a rich and extensive lineage, making it one of the oldest fields of human endeavor. As such, it warrants comprehensive exploration within the scope of this chapter. To fully grasp the breadth and application of welding, the subsequent sections will present an overview of its contemporary usage. In conclusion, the journey of welding from its rudimentary beginnings in medieval times to its modern form has been a testament to the collaborative efforts of countless researchers, scientists, and inventors. The early advancements made by figures like Sir Humphry Davy, August de Méritens, Nikolai Benardes, and others set the stage for the rapid growth and refinement of welding techniques. Today, welding has evolved into a sophisticated discipline with diverse applications across various industries, solidifying its position as an essential technology in the realm of metal fabrication and construction.

4.2 TYPES OF WELDING PROCESSES

As previously mentioned, welding is a crucial manufacturing method used for permanently joining parts together. These parts can be composed of various materials, including metals, ceramics, polymers, and composites. The welding process can be classified in several ways, which include:

1. **Melting or non-melting**: Welding techniques can involve either melting or non-melting processes. Melting welding methods, such as fusion welding, involve heating the materials to their melting point to achieve bonding. Non-melting techniques, such as solid-state welding, achieve connection without melting the materials.
2. **With or without pressure**: Welding processes can also be categorized based on the presence or absence of applied pressure during the joining process. Pressure-assisted welding techniques, such as friction welding, employ external force to enhance bonding. In contrast, non-pressure welding methods, like autogenous welding, achieve fusion without the application of external pressure.
3. **With or without filler**: Some welding processes utilize a filler material to facilitate bonding between the parts being joined. This filler material, typically a metal alloy, is melted to form a bridge between the base materials. Welding methods without filler, known as autogenous welding, rely solely on the fusion of the base materials for connection.

Creating a welding connection requires the application of energy. This energy can be generated and applied through various means. One method of classifying welding techniques is based on the source of the energy supply. These energy sources include electrical, chemical, mechanical, radiation, and sound. Study [1] illustrates a chart depicting different types of welding based on this classification. Given the wide scope of welding techniques, it is essential to highlight some common methods frequently utilized in various industries. The subsequent sections will provide an overview of these prevalent welding methods.

4.2.1 Oxyacetylene Welding

Oxyacetylene welding (OAW), also known as gas welding or flame welding, is a specialized form of oxygas welding that falls under the categories of fusion and chemical welding. Oxygas welding encompasses welding methods that utilize a combination of burning gas and oxygen as the heat source required for welding. In the case of OAW, the process involves the manual operation of a torch nozzle. The flame produced by the combustion of acetylene gas with oxygen at the tip of the torch nozzle generates sufficient heat to melt the metal of the workpiece. Once the molten metal solidifies, a connection is established within the workpiece. It is important to note that acetylene and oxygen are proportionally mixed in a burner mixing bag before being emitted through the nozzle and ignited [1].

Acetylene, a colorless hydrocarbon gas with the chemical formula C_2H_2, plays a critical role in OAW. Compared to other hydrocarbon gases, acetylene has a higher weight percentage of carbon. Additionally, acetylene possesses certain characteristics, such as an unpleasant odor and being lighter than air. The unpleasant smell can be attributed to impurities like hydrogen sulfide and hydrogen phosphorus. Notably, these gases are produced through the interaction of water with carbide rock.

In flame welding, the temperature directly affects the heat intensity and the welding outcome. The heat from the flame is transferred to the workpiece through both heat conduction and radiation. It is important to highlight that convective heat transfer depends on pressure, while radiative heat transfer depends on the fourth power of the absolute flame temperature, following the Stefan-Boltzmann law. Hence, even slight variations in the absolute temperature of the flame can result in significant changes in the amount of radiant heat [1].

Despite being an older welding method, OAW continues to find widespread use in various industries due to its notable advantages. This method is known for being easy to use, cost-effective, and portable. Study [1] presents a schematic representation of the equipment used in OAW.

In conclusion, OAW, also known as gas welding or flame welding, is a distinctive form of oxygas welding that relies on the combustion of acetylene gas with oxygen to produce the required heat for welding. The characteristics of acetylene, such as its higher carbon content and lighter-than-air properties, contribute to its role as a primary component in the welding process. OAW remains widely employed due to its ease of use, affordability, and portability, making it a valuable technique in the industry. Understanding the principles and equipment associated with OAW lays the foundation for further exploration of welding methods and their applications in modern manufacturing.

4.2.2 Arc Welding

One of the most widely used welding methods involves utilizing an electric arc as a primary heat source. This process relies on the passage of electric current through an ionized channel, which contains plasma and results in the generation of heat and light. The heat generated by the electric arc can be utilized for localized melting, facilitating the connection of workpieces, or melting the electrode or filler metal. Essentially, electrical energy is converted into thermal energy in this process. Such welding techniques are collectively referred to as "arc welding." Arc welding can be further categorized into two main groups based on the method employed to protect the molten pool: "slag protection" and "gas protection." Each category utilizes distinct mechanisms to shield the molten pool from atmospheric contamination and oxidation. These protection methods will be explored in greater detail in the subsequent sections.

4.2.2.1 Flux Shielded Arc Welding

If the molten pool is protected by slag, arc welding will be of the "slag protection" type. Shielded metal arc welding (SMAW) and submerged arc welding (SAW) are examples of this family, which will be explained below.

4.2.2.1.1 Shielded Metal Arc Welding

Electric arc welding with a covered electrode stands to be one of the most popular and widely utilized welding processes, often performed manually. This method involves the creation of an electric arc between a coated electrode and the workpiece, generating the necessary heat to melt both the base metal and the electrode. The coating on the electrode plays a crucial role in protecting the molten pool during welding.

During the welding process, the heat decomposes the electrode coating, which subsequently forms a protective shield in the form of slag and gas. The primary function of the coating is to safeguard the weld metal and prevent it from cooling too rapidly. The slag acts as a thermal barrier, slowing down the cooling process and enhancing the overall quality of the weld. Study [1] illustrates the physics schematic and the actual view of the process, respectively.

Electric arc welding with a covered electrode, also known as SMAW, involves several essential components. These components include the welding motor, electrode cable, electrode holder, electrode, workpiece clamp, workpiece cable, welding motor, and power supply. A closer observation reveals that the electrode and the workpiece are integral parts of the electric circuit. The circuit commences with the power supply, with one cable connected from the power source to the workpiece and the other to the electrode holder. Welding begins when an arc is established between the electrode tip and the workpiece. The heat generated by the arc melts the electrode tip and the adjacent surface of the workpiece. Consequently, small droplets of molten metal form rapidly at the electrode tip and are transferred to the molten pool through the arc current. Simultaneously, the filler metal is deposited as the electrode is gradually consumed. The arc progresses along the workpiece at a specific length and speed, melting a portion of the base metal and creating the resulting weld [1].

The SMAW process finds extensive usage across various industries due to its affordability, portability, and adaptability to diverse welding conditions. SMAW welding is particularly suitable for welding steel with a thickness exceeding 3 mm. Moreover, it offers versatility in welding a wide range of materials, including carbon steel, aluminum, stainless steel, and nickel [1].

The cost-effectiveness of SMAW welding makes it a preferred choice in industries where budget constraints are a consideration. Additionally, its portable nature enables welding operations to be conducted in various locations, including remote or challenging environments. The adaptability of SMAW to different welding conditions further enhances its practicality and applicability. The ability to weld a variety of materials expands the potential applications of SMAW across industries.

In summary, electric arc welding with a covered electrode, or SMAW, relies on a well-defined set of components to facilitate the welding process. The establishment of an arc between the electrode and the workpiece generates the necessary heat to melt the electrode and the adjacent workpiece surface. The resulting molten metal droplets contribute to the creation of the weld, while the electrode's gradual consumption allows for the deposition of filler metal. SMAW welding offers affordability, portability, and adaptability, making it suitable for welding thick steel and a diverse range of materials. Its utility in various industries is evident from its application in different welding conditions.

4.2.2.1.2 Submerged Arc Welding

The first SAW process, also known as "electric arc welding on the consumable wire under the powder," was officially registered in 1935. This technique involved the use of an electric arc beneath a layer of powder. It was primarily developed during World War II and gained prominence for its application in welding the T-34 tank. SAW is a process in which the necessary welding heat is generated by one or more arcs between the bare metal and one or more consumable electrodes. This method is referred to as "Submerged" because the electrode's tip is positioned within a bed of specialized mineral powder. The arc is formed beneath the powder and along the desired welding area, resulting in a successful weld. Consequently, the electric arc is not visible during the welding process. Part a of Figure 4.1 provides a schematic illustration of the physics involved in the SAW process.

Another distinctive feature of this welding method is the continuous feeding of the uncoated wire from the spool. By establishing an arc, the wire facilitates the connection at the welding site. In this process, the protection of the welding melt is achieved by the welding powder, which enters the workpiece surface through a channel from the holding tank. Part b of Figure 4.1 demonstrates the flow of electric current from the power source to the control system. Subsequently, the current, along with the wire supplied from the coil and power supply, enters the torch. Prior to the electric current, the powder is injected onto the workpiece

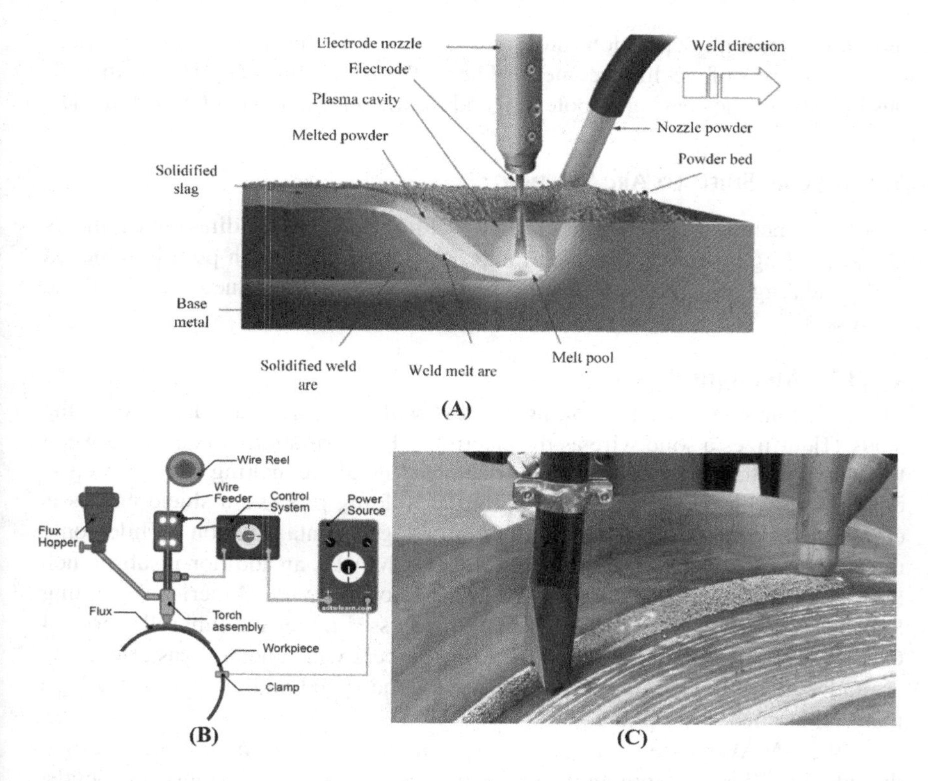

FIGURE 4.1 Submerged welding: (a) schematic view of the physics of the process, (b) schematic view of the equipment, and (c) actual view during the process.

surface, forming a layer of powder. The electric current and wire are then introduced onto the powder bed, creating a molten pool beneath it. It is important to note that throughout the entire process, the holding clamp must remain connected to the workpiece to establish the electric current [1].

The submerged welding method offers several advantages, including the minimal requirement of protective coating by the operator, the highest deposition rate among fusion welding processes (up to 18 kg/h), and efficient material utilization. Part c of Figure 4.1 illustrates the actual view of the SAW process for a better understanding. However, one limitation of this method is the constraint it imposes on welding geometry. Since the flux necessary for welding can easily flow beneath the powder, the welding is limited to cases where the flux can still remain beneath the powder and adequately protect the molten pool.

In summary, the SAW process, developed and registered in 1935, involves the use of an electric arc beneath a layer of powder to generate the necessary heat for welding. The uncoated wire continuously feeds from the spool, creating an arc and facilitating the welding connection. The welding powder protects the molten pool, and the process offers advantages such as reduced protective coating requirements, high deposition rates, and efficient material usage. However, it is limited in terms of welding geometry. The SAW process finds application in various industries, showcasing its potential and significance in the field of welding [1].

4.2.3 Gas-Shielded Arc Welding

Processes such as metal inert gas (MIG) welding and TIG welding are examples of arc welding processes in which the protection of the molten pool is achieved with shielding gas. In the following sections, we will introduce each of these processes.

4.2.3.1 Metal Inert Gas

The consumable electrode welding process with a neutral gas, depicted in the study [1], utilizes a solid wire as the electrode. The formation of plasma between the wire and the workpiece generates heat, leading to the melting and connection of the wire with the workpiece. During the welding process, a shielding gas is employed to protect the molten pool from oxygen contamination. While some refer to this process as gas metal arc welding (GMAW), an additional sub-branch known as metal active gas (MAG) exists. However, the American Welding Society categorizes both MIG and MAG processes under the umbrella term of GMAW. The consumable electrode welding process with a neutral gas offers several advantages, including ease of learning, good speed and quality, and a high material efficiency of up to 98%.

In the GMAW process, various gases such as argon, helium, nitrogen, carbon dioxide, and their combinations are employed to weld a wide range of metals, including aluminum, copper, stainless steel, steel, magnesium, and nickel alloys. Study [1] illustrates the actual view of this process applied in pipe welding. The equipment for the GMAW process consists of a welding engine, gas cylinder, wire feeding system, gun, and workpiece clamp. Parameters such as wire feeding rate, gun angle, voltage, and current intensity significantly influence the properties of the resulting weld. The operation of the GMAW process is similar to that of other electric arc processes.

4.2.3.2 Gas Tungsten Arc Welding (GTAW) or TIG

Tungsten electrode arc welding with a neutral shielding gas is a highly specialized and widely used arc welding process in the field. It is known by various names, such as GTAW and TIG. While this method was initially employed for welding airplanes in the 1930s, its widespread adoption occurred during the 1940s. A significant application of this process is in the welding of aluminum and magnesium parts, which were historically challenging and prone to porosity before the introduction of the TIG process [1].

As illustrated in the study [1], one pole of the current is connected to the workpiece, while the other pole is connected to a non-consumable electrode, typically made of tungsten. Since tungsten has a melting temperature of 3,410°C, it remains in a solid state during the welding process and serves the sole purpose of creating an arc with the workpiece. In TIG welding, a neutral gas, often argon, is utilized to shield the molten pool from environmental factors, such as oxidation caused by oxygen. Additionally, welding wire is employed as filler metal in this process.

4.2.4 Resistance Spot Welding

Resistance spot welding is a commonly employed form of resistance welding widely used in various industries. This process relies on generating heat through friction between two pieces of material. Specifically, the localized heat is produced due to the material's resistance to the passage of electricity. The connection is established by applying pressure using specialized equipment. The time required for the electric current to pass through the materials depends on several factors, including the type and thickness of the parts, the current intensity, and the shape of the electrodes' cross-section.

The AWS A3.0 standard recognizes resistance spot welding as both fusion welding and a solid-state welding process. This classification is based on whether melting occurs during the process. Depending on the amount of heat generated by the current passing through the base metal, there is a possibility of partially melting the welding point's surface or transforming it into a "paste-like" state. However, by applying appropriate pressure, a sound and reliable connection can be achieved. Thicknesses ranging from 0.001 to 25 mm can be accommodated using this process.

4.2.5 Ultrasonic Welding (USW)

Ultrasonic welding is a solid-state welding method commonly used to join plastic and metal parts, whether they are homogeneous or non-homogeneous. This process utilizes high-frequency oscillations to connect the parts together. Unlike other welding processes, ultrasonic welding does not require the use of powders or filler metals, and it does not involve the passage of electric current through the workpiece. Instead, it generates local heat within the welding area, which is below the melting temperature of the material. As a result, the workpiece remains in a solid state during the welding process.

During the ultrasonic welding process, the parts to be joined are positioned on a hard surface known as an anvil. The converter converts high-frequency energy (20–70 kHz) into oscillations, causing the tip of the horn or roller to oscillate. This generates transverse waves in the form of tension within the workpiece. These waves combine with the applied static force, resulting in an increase in dynamic shear stress within the workpiece. When the shear stress reaches a significant value, localized plastic deformation occurs at the contact point of the workpiece. Additionally, the oxidized coating on the surface of the workpiece is broken, allowing direct contact between the pure workpieces. Through this process, a metallurgical bond is achieved in the solid state without melting the base metal.

Although ultrasonic welding finds wide application in electronic connections, its conventional use involves joining leads to foil, sheet, aluminum wires, and connecting fine lead wires to each other. In study [1], the ultrasonic welding process consists of two main components: (1) an electrical stimulation source or frequency converter that transforms energy from a frequency of 60 Hz to high-frequency electrical energy, and (2) a converter that converts high-frequency electrical energy into mechanical oscillations. In a detailed analysis, the ultrasonic welding equipment can be further divided into the power source, converter, amplifier, and horn. The power supply increases the frequency of city electricity from 50–60 Hz to 20–50 kHz (or sometimes even up to 70 kHz). The converter then converts this electrical wave into high-frequency mechanical vibrations (ultrasonic waves). Subsequently, the waves produced by the converter are amplified, increasing their amplitude. The horn transmits these sound waves to the workpiece, exerting pressure to create a connection. When the sound waves reach the joint area, the high friction resulting from the molecular movement of the part surfaces converts the energy into heat, leading to the softening, melting, and creation of welding conditions.

4.2.6 Electron Beam Welding (EBW)

Electron beam welding is a widely employed melting process utilized in various industries, such as automotive, aerospace, electronics, and shipbuilding. The process involves the use of an electron beam welding machine to generate temperatures exceeding 100,000°C in the workpiece. At such high temperatures, metals and ceramics evaporate, leading to the formation of deep, solid, and distortion-free welds in both homogeneous and non-homogeneous parts.

The concept of metal melting using an electron beam was initially proposed by Crooks in 1879. Subsequently, Pirani made the first attempt to melt metals with an electron beam in 1907. The first electron beam welding was then introduced in the 1950s, utilizing a vacuum furnace. The electron beam process enables the welding of steel with thicknesses ranging from 0.01 to 250 mm and aluminum with a thickness of 500 mm, achieving an energy density of 12–10 W/m^2. The weldability of different materials using electron beam welding can be observed in a study [3]. It is important to note that the diameter of the electron beam typically ranges from 0.1 to 0.8 mm. Consequently, the resulting welds often exhibit a keyhole configuration, which will be further elaborated upon in the relevant chapter [3].

The electron beam welding process derives its energy from the rapid movement of focused electrons onto the workpiece. When the electrons strike the metal surface, they transfer almost all of their kinetic energy to the workpiece in the form of heat. Typically, this process is carried out in a vacuum environment of approximately 3–10 to 5–10 mm of mercury to minimize welding-related contamination within the vacuum furnace. Moreover, maintaining this level of vacuum is crucial for achieving a stable and uniform concentration of electrons on the workpiece surface. However, recent studies have also explored the possibility of conducting electron beam welding in non-vacuum conditions. The types of electron beam welding devices encompass high-pressure, low-pressure, and non-vacuum systems, though the intricacies of these variations extend beyond the scope of this book. Generally, most electron beam devices comprise an electron beam generator, an optical monitoring system, a workpiece holder, and a vacuum furnace [3].

The welding process within an electron gun, considered the core of the electron beam welding system, involves high-energy and high-speed electrons. These electrons are accelerated by passing through an electric field generated by a coil. In this state, the beam impacts the workpiece with substantial energy, resulting in the conversion of this energy into heat at the desired location on the workpiece. The heat causes the workpiece to melt, and upon solidification, welding is achieved within the workpiece.

4.2.7 Laser Beam Welding Process

As mentioned in the introduction chapter, laser beam welding is a modern fusion welding method that has gained significant attention in the industry in recent years due to its exceptional precision and speed. In this process, the surface of the workpiece is irradiated with a laser beam, generating the necessary heat to create a molten pool. Subsequently, upon solidification, the workpiece is locally welded. A schematic representation of the laser beam welding process, illustrating the stages before, during, and after welding, can be observed in the study [1]. Additionally, the study [4] provides an actual view of this process. It is worth noting that a notable advantage of laser welding, distinguishing it from other common fusion welding methods, is the production of thin and deep welds. This characteristic makes laser welding highly suitable for applications requiring a high ratio of penetration depth to weld width. As the subsequent chapters will delve into the details and various aspects of laser welding, the elaboration of additional specifics pertaining to this process will be avoided in this section [4].

4.2.8 Comparison of Different Welding Methods

In the previous parts, various welding processes were described. In the following, for a better understanding, some of them will be compared with each other from different points of view. In study [5], the energy density of various welding processes, including plasma, laser, Flux cored arc welding (FCAW), and GMAW, can be seen. It provides a valuable comparison of power density among different welding methods, namely plasma welding, GMAW, FSAW, and laser welding. The figure clearly demonstrates that plasma welding exhibits a higher power density than the other three methods. Moreover, it is noteworthy that both laser and electron beam welding processes exhibit the highest power density among all the methods. This high power density plays a crucial role in achieving optimal penetration depth.

Upon analyzing these studies, it becomes evident that when the laser beam is focused on the workpiece, both the laser and electron beam welding processes yield a greater penetration depth while maintaining a lower thermal area. This particular characteristic stands as a significant advantage of laser and electron beam welding when compared to other conventional welding processes [5].

4.3 WHY LASER BEAM WELDING?

Laser beam welding (LBW) is a highly favored process for material processing utilizing lasers. What makes laser welding so appealing? With other readily available welding methods such as MIG, TIG, and plasma welding in workshops, why does laser welding capture the attention of craftsmen and engineers? The attraction to laser welding stems from several significant advantages recognized by industry experts. These highlights are as follows:

- **Very high accuracy**: One of the most important advantages of laser welding, compared to electric arc and resistance methods, is its exceptional accuracy combined with low input heat. In some cases, laser welding requires the precise focus of a narrow laser beam on connections as thin as 0.02–0.05 mm in butt and corner welding. The accuracy of the laser beam's path depends on three variables: the precision of the robot's movement, the workpiece geometry, and the repeatability of the fixture holding the workpiece. By maintaining control and stability over these variables, laser welding achieves highly accurate welds [6].

  To enhance the accuracy of laser welding, techniques such as "seam tracking" play a crucial role, often based on imaging [7]. For example, a study employed seam tracking methods to trace the seam during laser welding. The study focused on welding two plates with a thickness of less than 2 mm, requiring the welding gap's width and deviation to be less than 0.1 mm. The diameter of the laser beam used in the process was approximately 0.3 mm. Precise movement of the laser beam along the center of the joint is vital for achieving optimal welding quality. Accurate measurements of the seam width and center are essential aspects of achieving high-quality welds [8,9]. In another study, the width and position of the weld seam in the transverse direction were controlled with an impressive accuracy of 0.006 and 0.008 mm, respectively, at a speed of 83 mm/s. The objective of this study, illustrated in study [10], was to ensure the welding path was precisely aligned with the desired welding location. For a detailed understanding of the algorithm and how the system operates, refer to reference [10].
- **The ability to create complex connections**: Another noteworthy advantage of laser welding is its ability to create complex connections that were previously unattainable with traditional welding methods. Laser welding opens up possibilities for joining intricate and three-dimensional parts. For instance, a study conducted by Huang et al. in 2013 explored the welding of a complex part with a narrow gap width of 0.1 mm using the welding seam tracking technique [7]. Study [7] illustrates the welding path on a 2-mm-thick curved aluminum workpiece during the laser welding process. To achieve this complex connection, the laser welding process employed a power of 3 kW and a speed of 58.33 mm/s.

The ability to weld complex parts with narrow gaps demonstrates the versatility and precision offered by laser welding.

- **Low heat-affected zone (HAZ)**: Another significant advantage of laser welding is its ability to create a low heat-affected zone. Due to the localized application of heat in the laser welding process, only a small area of the workpiece is affected by heat, resulting in a reduced HAZ. Consequently, the distortion and residual stress in the workpiece after welding are significantly minimized. This characteristic makes laser welding particularly suitable for applications where low distortion is critical, such as in the production of expensive and decorative products like jewelry.

  A comparative study conducted by Murad et al. in 2012 examined the thermal effective zone of lip welding using two methods: laser welding and GTAW [11]. The study involved a 6.4-mm-thick workpiece made of 2205 duplex stainless steel. Laser welding was performed with a maximum power of 8 kW, a welding speed of 8.3 mm/s, and argon shielding gas with a flow rate of 20 L/min. GTAW process parameters included a current of 110 A, a voltage of 12 V, a welding speed of 2.5 mm/s, and argon shielding gas with a flow rate of 15 L/min. Study [11] displays the upper section and the metallographic image of the GTAW-welded sample, respectively. Similarly, this study shows the upper section and the metallographic image of the weld obtained by the laser method. A comparison of study [12] reveals that the thermal effective area of the weld produced by the GTAW method is significantly larger than that of the laser method. These findings align with another study, which reported a thermal effective area of 3 mm for GTAW and 0.8 mm for laser welding, confirming the superior performance of laser welding in terms of minimizing the HAZ.
- **Creating a repeatable weld**: Laser welding is the preferred choice of manufacturers due to its ability to create repeatable welds in the workpiece. Compared to traditional welding methods, laser welding offers excellent welding quality with high repeatability, which ultimately reduces production costs. However, several conditions need to be considered to ensure reproducible welding.

  In the laser welding process, three key aspects must be controlled: the welding conditions, the operation of the welding device, and the execution method. To achieve this control, condition monitoring systems are employed during the welding process [13]. Conventional methods for monitoring these systems include the use of optical sensors such as visible, ultraviolet, and infrared cameras, audio monitoring, spectrometers, and pyrometers [14]. Study [15] illustrates an example of a monitoring system used in laser welding, consisting of a visible camera, an infrared camera, and a pyrometer [15]. For a more comprehensive understanding of monitoring systems, it is recommended to refer to reference [14]. By implementing condition monitoring systems and utilizing appropriate equipment, manufacturers can ensure the stability and consistency of the laser welding process. These monitoring systems provide real-time feedback and enable adjustments to be made during the welding process, resulting in reproducible and high-quality welds.

- **High joint strength**: The laser welding process offers the advantage of achieving high joint strength in manufactured products without the need for secondary fillers. It enables the creation of high-quality welds in slag-free conditions, which is particularly desirable in industries such as medicine.

  A study investigated the strength and mechanical properties of laser, TIG, and MIG welding in two-phase DP780 steel sheets with a thickness of 3.2 mm. The laser welding parameters included a power of 4 kW, a speed of 20 mm/s, and helium shielding gas with a flow rate of 20 L/min. TIG and MIG welding processes were performed with similar current and voltage settings, varying speeds, and shielding gases. Table 4.1 presents the mechanical properties of the joints resulting from the laser, TIG, and MIG welding methods, along with the properties of the base metal [16].

From Table 4.1, it is evident that the yield and ultimate strength of the laser welding joints are significantly higher compared to the TIG, MIG methods, and the base metal. Moreover, the laser welding method achieves this strength while exhibiting less elongation (23.7%) compared to the part before welding. Microscopic images of the fracture locations of the samples welded by the three methods—laser, TIG, and MIG—along with their stress-strain diagrams. The images reveal that the fracture location of the laser-welded sample has the greatest distance (6 mm) from the welding zone compared to the other methods. Additionally, the stress-strain diagram demonstrates that, for the same strain, the stress incurred by the laser welding method is higher than that of the other two methods. These observations confirm the superior strength of laser welding over conventional methods. Hence, it can be concluded that laser welding outperforms MIG and TIG methods in terms of strength, fracture location, and elongation [16].

TABLE 4.1
Mechanical Properties Resulting from Welding by Three Methods: Laser, TIG, and MIG Compared to the Workpiece [16]

Mechanical Characteristics→ Welding Process↓	Yield Strength (MPa)	Ultimate Tensile Strength (MPa)	Elongation (%)
Before welding (DP780)	566	800	25.1
Laser	621	816	23.7
TIG	612	778	19.9
MIG	537	692	21.1

4.4 A BRIEF HISTORY OF LASER BEAM WELDING APPLICATIONS

The laser welding process has found numerous applications across different industries. Its initial use can be traced back to the 1970s, when it was employed in jewelry making. Subsequently, the process gained popularity in various medical techniques [17]. The significant advancements and widespread adoption of laser welding can be attributed to the introduction of high-power CO_2 laser devices in countries like America, Japan, Germany, England, and France around 1990. For instance, in Japan, three 45 kW CO_2 lasers and two 10 kW YAG lasers were introduced in 2003. The market also witnessed the introduction of disc, fiber, and diode lasers, further expanding the practical application of laser welding [18]. Given the extensive reach of laser welding, it is worth highlighting some of the diverse fields where this process is employed. The following section provides an overview of these fields, showcasing the versatility and utility of laser welding.

4.4.1 Laser Welding in Dentistry

The application of laser welding in dentistry encompasses two main areas: "Intraoral" and "Extraoral". The use of laser welding in the intraoral field, specifically referred to as "Intraoral Laser Welding" or ILW, has gained significant attention among researchers. The primary objective of employing laser welding in this domain is to facilitate the connection and repair of dental prostheses, artificial teeth, and, in some cases, even tooth repairs. The utilization of laser welding techniques within the field of dentistry holds great potential for enhancing the precision and effectiveness of dental procedures. Researchers are actively investigating and exploring the applications of laser welding in various dental applications to further improve the quality of dental treatments and prosthetic restorations.

In the field of dentistry, various lasers are utilized, including CO_2 lasers with a wavelength of 1,060 nm, diode lasers with a wavelength of 810 nm, and Nd:YAG lasers with a wavelength of 1,064 nm. According to several studies, Nd:YAG lasers are considered more suitable for dental welding compared to other types of lasers. This preference is due to the following reasons:

1. The pulse interval of dental CO_2 lasers is very short, typically in the range of microseconds, which does not provide sufficient heat for metal melting.
2. Diode lasers have relatively low output power, ranging from 5 to 10 W, which is insufficient to initiate the welding process.

In a particular study, as depicted in study [17], an Er:YAG laser with a wavelength of 2,940 nm and an Nd:YAG laser with a wavelength of 1,064 nm, capable of adjusting to two different wavelengths, were used as laser sources [17]. This study illustrates the head used in this research, which had a beam diameter of 2 mm.

TABLE 4.2
Chemical Structure of Alloys Used in Dental Laser Welding

Weight Structure (%)→ Alloy↓	Ni	Co	Cr	Mo	Si	Other	References
G-Mou	67	–	26	5	1.5	C, B, Mn <1	[15]
Remanium Cs	61	–	26	11	1.5	Fe, Ce, Al <1	[15]
GM-800	–	64.5	30	5	–	C:0.3, Mn: 0.2	[15]
Nichrominox	–	–	18.26	0.32	0.41	Cu:0.34, Mo:0.32, S:0.001, Mi:11.22, C: 0.018	[16]

Various alloys find applications in laser welding for dentistry. Titanium and its alloys have gained prominence in recent years due to their favorable biostructural and superior mechanical properties, making them suitable for the production of bridges, crowns, and artificial teeth [19]. Table 4.2 provides an overview of other alloys commonly used in dentistry. For instance, inexpensive G-Mou alloy is employed in dental bridges and crowns, while Remanium Cs alloy is utilized in fixed ceramic dentures. GM-800, a cobalt-chromium alloy, is employed for both fixed and removable dentures within the oral cavity [20]. Additionally, French stainless steel wires called Nichrominox, with a diameter of 0.8 mm, are used in orthodontics [21]. To gain a better understanding of the application of laser welding in both extra-oral and intra-oral areas, examples will be presented in Sections 4.3.1.1 and 4.3.1.2, respectively.

4.4.1.1 Extraoral Laser Welding

Study [17] showcases an orthodontic mold used for an 8-year-old boy during a routine dental examination. In this study, a broken section can be observed in the movable orthodontic appliance. The damaged area was initially repaired using laser welding without the need for filler metal, followed by the application of a plastic membrane to cover the repaired portion. Notably, the high speed and precision of laser welding facilitated the efficient repair of the orthodontic device within a few minutes, allowing it to be placed back in the patient's mouth [17]. It is important to mention that orthodontic wires are typically made of stainless steel with a diameter of 0.8 mm. Furthermore, the laser welding conditions utilized for repairing this particular orthodontic appliance were as follows: 800 W of power, a pulse time of 3.6 ms, an energy level of 2.8 J, and a focal distance of 0.90 mm [21].

4.4.1.2 Intraoral Laser Welding

In one instance, a 45-year-old patient had a fractured implant located between two front teeth. Laser welding was employed to repair the middle bridge of the implant. To safeguard the surrounding oral tissues from potential laser beam damage, a cylindrical chamber was used during the welding procedure. Following the welding process, a layer of resin was applied to preserve the tooth's aesthetic appearance. The laser welding process was completed in approximately 7 minutes, and the patient experienced no discomfort throughout. Subsequent examinations conducted at 1, 2, and 6 months revealed no signs of defects or issues at the welding site within the patient's mouth [17].

To further explore intraoral laser welding, let's examine tooth-base welding. Typically, the base of the tooth is composed of pure titanium known as CP-Ti, as depicted in the study [22] in its healthy state. As shown in the study [22], an I-shaped incision was made at the center of the specimen. The incision area was thoroughly cleansed using the ultrasonic method and subsequently treated with aluminum oxide sandblasting. For comparison, separate welds were performed using the TIG and laser methods. The TIG welding parameters were set at a current of 36 A, a duration of 60 ms, and an electrode diameter of 1.6 mm. Laser welding, on the other hand, was carried out using a German Dentrum machine with an Nd:YAG source operating at a wavelength of 1,064 nm. The laser settings were 390 V/ms with a beam diameter of 0.2 mm. It is important to note that the voltage, amount, and time are parameters that control the size and duration of the energy pulse emitted by the laser device. The study revealed that laser welding resulted in a significantly more precise and higher quality connection compared to the TIG method, exhibiting reduced strain [22].

4.4.2 Automotive

Volkswagen, a prominent German car manufacturer, played a pioneering role in the utilization of laser welding within the automotive industry. Presently, Volkswagen employs laser welding for 75% of the body welding in Golf models, 35% in Polo models, and 50% in the latest Turan and Torg models. Notably, in the production line of Golf cars alone, the company has incorporated 250 laser welding machines, the majority of which are 4 kW Yag lasers manufactured by Trumpf. Each laser source provides laser energy to six welding heads mounted on robotic arms. The progress of laser welding technology has also extended to other automotive companies, including BMW (all series five, six, and seven cars) and Audi AJ cars (steel doors of Audi A3, A4, and Q5, and aluminum doors of Audi A6 and A8) [23].

For instance, the aluminum doors of the Audi A6, A7, and A8 series employ corner and overlapping welds. The adoption of laser welding technology in the automotive industry has facilitated enhanced manufacturing processes and improved structural integrity.

The car door is subjected to laser welding using 6000-grade aluminum along with AlMg4.5MnZr filler wire measuring 1.2 mm in diameter. The laser welding process employs a power of 3.6 kW and a speed of 16.67 mm/s. It is noteworthy that Audi's laser welding procedure typically encompasses a power range of 4–20 kW, a travel speed ranging from 2 to 5 m/min, and employs a filler wire with a standard diameter of 0.6 mm. Laser welding plays a pivotal role in ensuring the structural integrity and durability of car components, such as doors. Audi's utilization of laser welding technology in their manufacturing processes showcases its significance in achieving precise and reliable welds.

4.4.3 Railway

Laser welding is highly recommended for applications that demand fast welding with minimal thermal distortion, especially in thin weld seams. The railway industry has embraced laser welding, and Photon AG, a German company, has been at the forefront of its implementation. According to Photon AG, the utilization of laser welding technology has led to a notable increase in the speed of manufacturing train components, achieving a two- to three-fold improvement. The precision achieved through laser welding is unparalleled, resulting in reduced re-welding costs. Furthermore, laser welding enhances process flexibility, allowing for automated control and customization based on customer requirements. This integration of laser welding technology has revolutionized the railway industry, offering enhanced efficiency, precision, and cost-effectiveness [24].

To investigate further of laser welding in the railway industry, it is beneficial to examine an industrial example within this field. A notable project in Europe, known as ICx, involved a contract for 160 trains valued at approximately 4 billion euros between 2017 and 2023. Photon Company played a significant role in this project by utilizing laser welding technology to weld the walls adjacent to the cabins. In total, 30 different parts were welded, comprising over 2,400 paths. Each 12-car train consisted of approximately 170 side panels, resulting in approximately 8,000 m of welding, all performed using a 6-kW laser source manufactured by Trumpf Company [24].

The production and construction of train cars involve the use of various materials, with steel and aluminum being the most commonly utilized metals. Ferritic steel, containing 12% chromium, is often employed for its exceptional corrosion and wear resistance. Aluminum, on the other hand, is widely used in the body of trains due to its low density (one-third of steel's density) and impressive strength (over half that of steel). Furthermore, aluminum exhibits remarkable corrosion resistance. For more than 50 years, aluminum has been extensively employed in wagon body construction. Different aluminum grades are employed, including five Al-Mg grades (e.g., 5005, 5754, and 5083), six Al-Mg-Si grades (e.g., 6005, 6060, 6061, and 6063), and seven Al-Zn grades (e.g., 7075), as examples of aluminum grades used in train bodies [25]. The characteristics of aluminum alloys, such as their high heat transfer coefficient and light reflectance, necessitate a welding method that concentrates input heat on the workpiece and enables rapid welding. The hybrid laser welding method, which combines laser welding with other techniques such as MIG, TIG, or plasma, fulfills these requirements [26].

In a study focusing on laser-MIG hybrid welding, the A6N01S-T5 grade of aluminum, known for its high extrudability, was utilized. This specific aluminum alloy allows for the production of components and hollow parts for wagon bodies. Study [26] illustrates a schematic view of the laser-MIG hybrid welding process and the weld joint in the hollow body of a train [26].

To validate the results, three types of weld connections were performed: MIG welding (standard crossbar mode) and MIG-laser hybrid welding (narrow crossbar and standard crossbar modes). It should be noted that the narrow standard saddle differs in dimensions, and further details can be found in reference [26]. The microhardness results, as depicted in part a of the study [26], indicated that both laser-MIG hybrid welding modes exhibited higher hardness compared to the MIG method. Additionally, the tensile test results revealed significantly higher tensile strength in the samples of both laser-MIG hybrid modes when compared to the MIG method. Therefore, it can be concluded that laser welding not only offers general advantages such as high speed and low distortion but also enhances physical properties such as strength when compared to conventional methods [26]. Similar results have been confirmed in other studies as well [27].

4.4.4 Shipbuilding

The first European study on laser welding for shipbuilding began in 1980 with the aim of utilizing this technology in the industry. The initial application of laser welding in shipbuilding, following previous research, took place in 1994 using CO_2 lasers [28]. The shipbuilding industry was among the early adopters of high-power and hybrid laser welding due to its ability to minimize distortion. The marine transportation sector recognized several advantages of laser welding, including high welding speed, minimal distortion, the capability to weld thick parts in a single pass, easy automation, and minimal adverse environmental effects. However, the primary advantage that drove the shipbuilding industry to adopt laser welding was the need to minimize distortion. Prior to the introduction of laser welding, approximately 20%–30% of labor time was dedicated to repairing distortion issues associated with conventional welding processes. Apart from geometric concerns, the mechanical properties of the welds were also a crucial consideration due to the joints being subjected to forces induced by strong ocean waves. Some of the problems and limitations of conventional welding methods in the shipbuilding industry include [29]:

- Limited malleability
  - Limited strength (in the face of shock and quasi-static forces)
- Cracks caused by environmental conditions (hydrogen cracking and cracking)
- Fatigue

Laser welding technology has found extensive application in the construction of cruise and passenger ships. One notable shipbuilding company that employs laser welding technology is Meer Vert, based in Germany. The company operates a laser center dedicated to ship construction, which was established in 2001 and utilizes four 12 kW carbon dioxide lasers. It is noteworthy that the construction of a cruise ship using 3 × 10-m plates requires approximately 1,000 km of welding, with laser technology accounting for approximately 800 km of that total. To grasp the scale of a cruise ship constructed by Meer Vert, let us consider the specifications of their largest production, Discovery of Adventure, in 2019. This ship spans 236 m in length, 31.2 m in width, and weighs 58,250 Gigatons. It boasts 15 decks, an engine power of 21.6 MW, accommodates 999 passengers in 554 cabins, and reaches a maximum speed of 18 km/h [29].

After discussing the generalities and applications of laser welding in the shipbuilding industry, let us delve into some specific details of this field. The construction of ship decks requires 20-m-long plates with varying thicknesses ranging from 4.5 to 30 mm. However, such plates are practically unavailable. To address this challenge and optimize consumables while meeting the required strength and geometry criteria, the concept of butt-welded sandwich panels was introduced. These sandwich panels consisted of 20-m-long plate strips that were connected to form a 20-m by 20-m deck using butt welding. Study [29] provides a visual representation of sandwich panels with butt welds. This approach effectively solved the problem of obtaining 20-m-long plates with variable thicknesses while ensuring strength, geometry, and minimal distortion, thanks to the utilization of laser welding technology [29].

In 2010, significant advancements were made at the Mirort laser center with the introduction of disk lasers. This technological upgrade enabled the construction of ship decks up to 30 m wide and 25 m long, with a maximum area of 750 m^2. The company's enthusiasm for laser welding was evident, leading to further enhancements at the laser center. Presently, the laser center has the capability to laser weld structures with an impressive area of 2,300 m^2, a length of 43.5 m, and a weight of 73 t. Remarkably, the company laser welds approximately 1,300 km of material each year, with a focus on low-thickness plates ranging from 5 to 8 mm.

4.4.5 Aerospace

The aerospace industry has faced limitations in the application of welding processes due to two main reasons, as supported by academic research:

- **Cracking**: The presence of cracks in welded connections poses a significant risk to the structural integrity of aerospace components [30]. This issue highlights the need for a welding process that can ensure high reliability and minimize the occurrence of cracks.
- **Aluminum alloy weldability**: Aluminum alloys, particularly grade 2 and grade 7, exhibit low weldability in fusion welding processes commonly used in aircraft bodies [31]. This low weldability extends to fuel tanks for projectiles, shuttles, and spacecraft [32]. Consequently, alternative welding methods with improved efficiency and effectiveness, such as laser welding, are in demand for aerospace structures.

To illustrate the advantages of laser welding in comparison to other welding methods, the study [33] provides a visual comparison based on the physical properties of welding, production rate, and cost. These figures demonstrate the potential of laser welding to address the challenges faced by the aerospace industry.

The aircraft industry, being one of the most significant consumers of aircraft structures, is particularly concerned with enhancing the metal body of airplanes to achieve an integrated structure while minimizing weight and construction costs. It is worth noting that approximately 55% of the total cost of an aircraft is allocated to the fuselage, which highlights the industry's focus on this aspect [30].

Traditionally, the aluminum body of an aircraft is constructed using panels with stringers. These panels, placed horizontally and longitudinally on the stringers and frames, form the aircraft's shell. Prior to the adoption of laser welding, the connection between stringers, frames, and panels was achieved through riveting, as depicted in the study [34], which provide a schematic view and location of rivets [34].

However, the riveting process presents several challenges, including low yield, limited upgradability, and high cost. To address these issues, a suitable solution emerged in the form of two-way fiber laser welding technology, as illustrated in the study [34]. This technology was initially proposed by Airbus as a replacement for riveted T-shaped connections [35]. By eliminating connections such as fasteners and seals, the weight of the structure is reduced, consequently leading to cost savings through process automation [36]. The speed of laser welding employed in this context is approximately 12 m/min, whereas riveting typically ranges between 0.15 and 0.25 m/min [37].

One might question why laser welding was chosen over other welding methods. The answer lies in the fact that alternative fusion welding techniques tend to introduce residual stresses and distortions, which can contribute to crack formation and propagation at the weld site during fatigue processes. In contrast, laser welding of aircraft aluminum structures generates lower residual stress and distortion levels [38]. Study [39] demonstrates the double-sided fiber laser welding process for panels and its positioning within the aircraft. Overall, the adoption of laser welding in the aircraft industry provides numerous advantages in terms of weight reduction, cost efficiency, and improved structural integrity compared to traditional riveting and other fusion welding methods.

Further information regarding the two-way fiber laser welding process in T-shaped joints, specifically using aluminum-lithium alloy 2060-T8/2099-T83 for aircraft body panels, was investigated in a study. The actual and schematic views of this process can be observed in the study [35]. The process parameters included a power of 3 kW, a speed of 10 m/min, a wire feed rate of 4.3 m/min, a beam angle of 22°, wire injection and gas injection angles of 20°, argon shielding gas with a flow rate of 15 L/min, a beam diameter of 0.26 mm, and a wire diameter of 1.2 mm. Comparative analysis revealed that the use of CW3 filler metal, with a chemical structure of Al-6.2Cu-5.4Si, resulted in reduced porosity and increased strength compared to the conventional filler metal AA4047, which has a chemical structure of Al-12Si. The cross-section of the weld produced by the two filler metals can be observed in the study [35].

In the production of aluminum-grade 6-stringer panels for Airbus aircraft, double-sided fiber laser welding has been employed. In specific models such as A318, A380, and A340, the two-way fiber laser technology has been used for welding two panels with over 50 m of welding, eight panels with over 300 m of welding, and fourteen panels with over 400 m of welding. By 2010, over 1,200

TABLE 4.3
Reducing Aircraft Production Costs While Reducing the Number of Panels Used by Using Laser Welding Technology [31]

Numbers of Panels→ Parameters↓	Base (10)	4	6	8
Engineering cost (%)	0	−30	−20	−10
Raw material cost (%)	0	−15	−10	−5
Manufacturing cost (%)	0	−50	−35	−20
Assembly cost (%)	0	−50	−30	−15
Weight (%)	0	−6	−4	−2

sheets were laser welded for Airbus. The substantial adoption of this technology can be attributed to several factors. One significant factor is weight reduction. By replacing rivets with laser welding, there is a reduction of 180 g/m of connection. Consequently, in the construction of A318, A380, and A340 aircraft, this leads to weight reductions of 9, 54, and 72 kg, respectively. While these individual weight reductions may appear small, laser welding contributes to an overall weight reduction of approximately 10% for each panel [31].

Another advantage of laser welding is its capability to produce reinforced and robust panels, facilitating the production of large-sized panels for easier final assembly of the aircraft body. Study [31] illustrates the circular section of the airplane body, which consists of multiple panels (typically ten panels) that are ultimately connected together. With laser welding, the number of panels can be reduced to eight or even four, as depicted in the study [31]. In addition to reducing the number of panels, this approach also reduces costs associated with engineering, raw materials, component manufacturing, assembly, and weight, as outlined in Table 4.3 [31].

The range of materials used in aircraft is different due to the variety of required properties. Because one or more materials alone cannot satisfy all the required properties of aerial structures. Therefore, a combination of different materials is used with the aim of balancing cost, performance, and safety in the structure, components, and body of the aircraft. In studies [30,40], some materials used in the production of airplanes are collected along with their properties and applications.

According to studies [30,40], the cost and properties used in the structure of aerial structures are very diverse. For example, aluminum and steel are the cheapest, composites are light weight, and steel has the highest strength. In this way, it can be seen that in the construction of an aerial structure, a combination of different materials is used to ensure the desired properties of each part.

4.4.6 Heat Transfer Systems

Heat transfer between streams with different temperatures has numerous applications in industrial, commercial, and domestic settings. One common method employed for this purpose is the use of heat exchangers. Heat exchangers are not a new technology and can be observed in natural systems as well. For instance, the transfer of heat between the external air and the blood in the lungs and air sacs of the human body serves as a clear example of a heat exchanger. Similarly, birds have a clever arrangement in their feet, resembling a heat exchanger, where veins transfer heat in the opposite direction. This mechanism allows birds to place their feet in cold rivers or streams without transferring the cold to their bodies [41]. Therefore, heat exchangers can also be found in biological organisms.

In the industrial sector, heat exchangers gained prominence during the industrial revolution. These devices come in a wide range of shapes, sizes, transmission mechanisms, and other features and are utilized in various industries such as electricity and power generation, oil processing, transportation, air conditioning, and cooling and refrigeration [42]. Heat exchangers can be classified based on different criteria, including the transfer process, number of fluids involved, construction, and fluid flow arrangement [43]. Study [44] depicts a visual representation of the classification of heat transfer systems based on their construction and applications.

The study [44] demonstrates the existence of different types of heat exchangers. Among them, plate heat exchangers have garnered significant attention in recent years within the industry due to their numerous advantages. These advantages include high efficiency and heat transfer coefficients, compact size and reduced weight, lower cost, ease of service and maintenance, and minimal sediment formation [45]. Study [44] depicts the wide-ranging applications of plate heat exchangers across industries such as food, ventilation systems, marine, petrochemical, and oil and gas.

However, plate heat exchangers face several limitations, including washer corrosion, restricted working pressure and temperature due to washer usage, limited washer lifespan, and the need for washer replacement [43]. To overcome these limitations, the construction of new-generation plate heat exchangers has shifted towards the utilization of laser welding technology. Laser welding is extensively employed in the fabrication of semi-welded, fully welded, and brazed plate heat exchangers. Studies indicate that the temperature range and working pressure for welded plate heat exchangers are –50°C to 350°C and zero to 40 bar, respectively [46]. Additionally, the use of welding in heat exchanger plates enhances their strength, thereby improving the efficiency of these systems under high temperatures and working pressures.

In part a of Figure 4.2, a schematic view of a semi-welded plate heat exchanger is presented. This heat exchanger is comprised of several components, including the fixed rod, front and rear frames, plates, base, and supporting bolts. Part b of Figure 4.2 provides detailed insight into the structure of the plate, gasket, and their respective welding locations. Semi-welded heat exchangers are constructed using corrugated plates that are welded together on one side, while the opposite side is sealed with a gasket (part c of Figure 4.2). These converters offer a key advantage in terms of easy disassembly, allowing for straightforward maintenance and cleaning procedures. Additionally, their capacity can be expanded

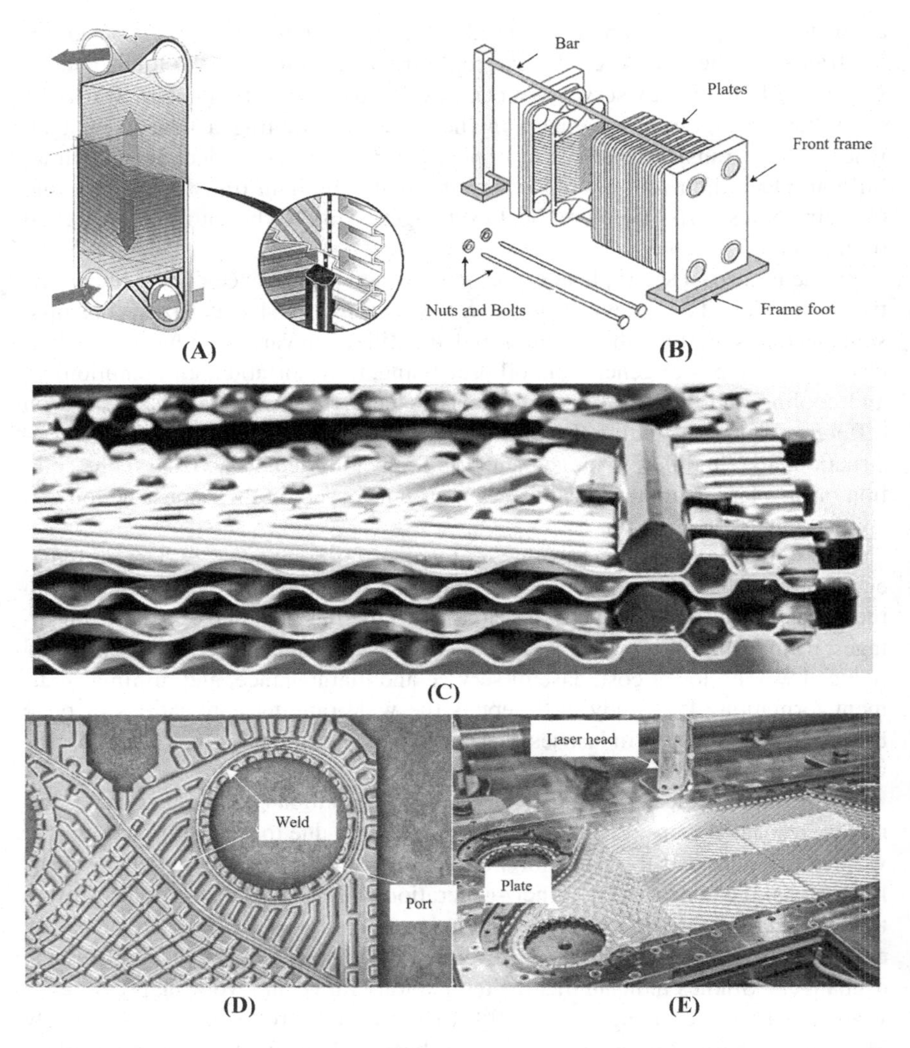

FIGURE 4.2 Semi-welded plate heat exchanger: (a) schematic of components, (b) schematic of plate, (c) actual view of welded plate and gasket, (d) laser welding process of plate, and (e) welded plate.

by increasing the number of plates, providing flexibility in meeting varying heat transfer requirements.

The working temperature range of semi-welded plate heat exchangers extends from –47°C to 170°C, withstanding pressures up to 40 bar (Figure 4.3). These converters find applications in processes involving condensation, water vapor, and refrigeration cycles. Laser welding plays a crucial role in enabling these applications and reaping the associated benefits. Parts d and e of Figure 4.2 illustrate the laser welding process of the plates and the resulting welded plate, respectively.

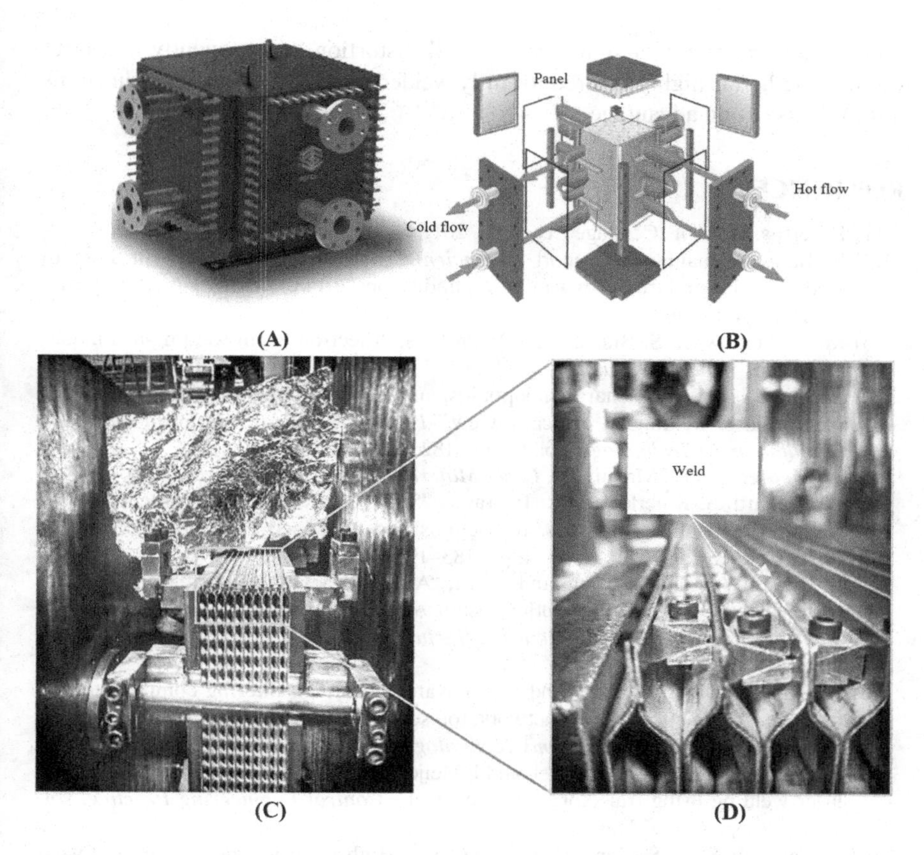

FIGURE 4.3 Welded plate heat exchanger: (a) schematic and (b) actual core, (c) weld seam, and (d) welded core.

It is important to note that due to the low thickness of the plates and the requirement for circular welding at the ports, laser welding is the only viable method for welding the plates of semi-welded heat exchangers. The utilization of laser welding in plate heat exchangers is driven by advantages such as the capability for complex welding and minimal distortion, particularly in thin plates.

Fully welded plate heat exchangers represent another type of plate heat exchanger. These converters are composed of various components, including upper and lower heads, plate packs, combs, baffles, washers, and panels. In fully welded plate heat exchangers, the flow patterns of the fluids move in the same direction within the corrugated plates. Baffles are employed to create single-path and multi-path flow configurations within the converter. These fully welded plate heat exchangers are designed to withstand pressures of up to 40 bar and working temperatures of up to 300°C. It is worth emphasizing that minimal distortion is of utmost importance during the welding of the plates in fully welded converters. Since the plates are placed near each other, any distortion in each plate can accumulate and result in significant overall distortion. Consequently, the laser welding method is highly favored for fully welded plate heat exchangers due to its minimal distortion advantage.

REFERENCES

[1] L. Jeffus, *Welding*. Cengage Learning, 2011.
[2] R. Balasubramaniam, A report on *Ancient Science and Technology* in Story of Delhi Iron Pillar. https://www.mysteryofindia.com/2014/08/mystery-of-1600-years-old-iron-pillar.html
[3] M. S. Węglowski, S. Błacha, and A. Phillips, "Electron beam welding-techniques and trends-review," *Vacuum*, vol. 130, pp. 72–92, 2016.
[4] J. Stavridis, A. Papacharalampopoulos, and P. Stavropoulos, "Quality assessment in laser welding: A critical review," *The International Journal of Advanced Manufacturing Technology*, vol. 94, pp. 1825–1847, 2018.
[5] W. M. Steen and J. Mazumder, *Laser Material Processing*. Springer, 2010.
[6] B. Regaard, S. Kaierle, and R. Poprawe, "Seam-tracking for high precision laser welding applications-Methods, restrictions and enhanced concepts," *Journal of Laser Applications*, vol. 21, no. 4, pp. 183–195, 2009.
[7] Y. Huang, Y. Xiao, P. Wang, and M. Li, "A seam-tracking laser welding platform with 3D and 2D visual information fusion vision sensor system," *The International Journal of Advanced Manufacturing Technology*, vol. 67, no. 1–4, pp. 415–426, 2013.
[8] X. Gao, X. Zhong, D. You, and S. Katayama, "Kalman filtering compensated by radial basis function neural network for seam tracking of laser welding," *IEEE Transactions on Control Systems Technology*, vol. 21, no. 5, pp. 1916–1923, 2012.
[9] M. De Graaf, R. Aarts, B. Jonker, and J. Meijer, "Real-time seam tracking for robotic laser welding using trajectory-based control," *Control Engineering Practice*, vol. 18, no. 8, pp. 944–953, 2010.
[10] P. Wang, W. Shao, S. Gong, P. Jia, and G. Li, "High-precision measurement of weld seam based on narrow depth of field lens in laser welding," *Science and Technology of Welding and Joining*, vol. 21, no. 4, pp. 267–274, 2016.
[11] A. I. Mourad, A. Khourshid, and T. Sharef, "Gas tungsten arc and laser beam welding processes effects on duplex stainless steel 2205 properties," *Materials Science and Engineering*: *A*, vol. 549, pp. 105–113, 2012.
[12] T. Pasang *et al.*, "Comparison of Ti-5Al-5V-5Mo-3Cr welds performed by laser beam, electron beam and gas tungsten arc welding," *Procedia Engineering*, vol. 63, pp. 397–404, 2013.
[13] C. Dawes, *Laser Welding: A Practical Guide*. Woodhead Publishing, 1992.
[14] D. You, X. Gao, and S. Katayama, "Review of laser welding monitoring," *Science and Technology of Welding and Joining*, vol. 19, no. 3, pp. 181–201, 2014.
[15] Y. Saadlaoui, J. Sijobert, M. Doubenskaia, P. Bertrand, E. Feulvarch, and J.-M. Bergheau, "Experimental study of thermomechanical processes: Laser welding and melting of a powder bed," *Crystals*, vol. 10, no. 4, p. 246, 2020.
[16] J. H. Lee, S. H. Park, H. S. Kwon, G. S. Kim, and C. S. Lee, "Laser, tungsten inert gas, and metal active gas welding of DP780 steel: Comparison of hardness, tensile properties and fatigue resistance," *Materials & Design*, vol. 64, pp. 559–565, 2014.
[17] X. Na, *Laser Welding*. BoD-Books on Demand, 2010.
[18] S. Katayama, *Fundamentals and Details of Laser Welding*. Springer, 2020.

[19] K. Iwasaki, S. Ohkawa, M. Uo, T. Akasaka, and F. Watari, "Laser welding of titanium and dental precious alloys," *Materials Transactions*, vol. 45, no. 4, pp. 1140–1146, 2004.
[20] C. Bertrand, Y. Le Petitcorps, L. Albingre, and V. Dupuis, "The laser welding technique applied to the non precious dental alloys procedure and results," *British Dental Journal*, vol. 190, no. 5, pp. 255–257, 2001.
[21] C. Bertrand, Y. Le Petitcorps, L. Albingre, and V. Dupuis, "Optimization of operator and physical parameters for laser welding of dental materials," *British Dental Journal*, vol. 196, no. 7, pp. 413–418, 2004.
[22] S. A. Rodrigues, A. G. C. Presotto, V. A. R. Barão, R. L. X. Consani, M. A. A. Nóbilo, and M. F. Mesquita, "The role of welding techniques in the biomechanical behavior of implant-supported prostheses," *Materials Science and Engineering: C*, vol. 78, pp. 435–442, 2017.
[23] M. Graudenz and M. Baur, "Applications of laser welding in the automotive industry," In: *Handbook of Laser Welding Technologies*. Elsevier, 2013, pp. 555–574.
[24] F. P. Incropera, D. P. DeWitt, T. L. Bergman, and A. S. Lavine, *Fundamentals of Heat and Mass Transfer*. Wiley, 1996.
[25] G. Posch, V. Holtsinger, and S. Bychkovskii, "Welding of railway wagons: Tasks in the area of materials, processes and automation," *Welding International*, vol. 29, no. 3, pp. 213–218, 2015.
[26] H. Yonetani, "Laser-MIG hybrid welding to aluminium alloy carbody shell for railway vehicles," *Welding International*, vol. 22, no. 10, pp. 701–704, 2008.
[27] S. Matteï, D. Grevey, A. Mathieu, and L. Kirchner, "Using infrared thermography in order to compare laser and hybrid (laser+ MIG) welding processes," *Optics & Laser Technology*, vol. 41, no. 6, pp. 665–670, 2009.
[28] G. Turichin, M. Kuznetsov, I. Tsibulskiy, and A. Firsova, "Hybrid laser-arc welding of the high-strength shipbuilding steels: Equipment and technology," *Physics Procedia*, vol. 89, pp. 156–163, 2017.
[29] S. Katayama, *Handbook of Laser Welding Technologies*. Elsevier, 2013.
[30] A. Zavadski, "Advanced welding technologies used in aerospace industry," 2018.
[31] P. M. Moreira, L. F. da Silva, and P. M. de Castro, *Structural Connections for Lightweight Metallic Structures*. Springer, 2012.
[32] G. Wang, Y. Zhao, and Y. Hao, "Friction stir welding of high-strength aerospace aluminum alloy and application in rocket tank manufacturing," *Journal of Materials Science & Technology*, vol. 34, no. 1, pp. 73–91, 2018.
[33] P. F. Mendez and T. W. Eagar, "New trends in welding in the aeronautic industry," In: *2nd Conference of New Manufacturing Trends*, Bilboa, Spain, 2002, pp. 19–20.
[34] S. M. Tavares, *Design and Advanced Manufacturing of Aircraft Structures Using Friction Stir Welding*. Universidade do Porto (Portugal), 2011.
[35] B. Han, W. Tao, Y. Chen, and H. Li, "Double-sided laser beam welded T-joints for aluminum-lithium alloy aircraft fuselage panels: Effects of filler elements on microstructure and mechanical properties," *Optics & Laser Technology*, vol. 93, pp. 99–108, 2017.
[36] Y. Chen, L. Li, W. Tao, and Z. Yang, "Laser welding technologies for aircraft fuselage panels," Conference: *Laser and Tera-Hertz Science and Technology*. Optica Publishing Group, November 2012, p. MTh5B, 2012.
[37] X. Sun, E. Shehab, and J. Mehnen, "Knowledge modelling for laser beam welding in the aircraft industry," *International Journal of Advanced Manufacturing Technology*, vol. 66, 2013.**
[38] F. S. Bayraktar, *Analysis of Residual Stress and Fatigue Crack Propagation Behaviour in Laser Welded Aerospace Aluminium T-Joints*. Technische Universität Hamburg, 2011.

[39] E. Schubert, M. Klassen, I. Zerner, C. Walz, and G. Sepold, "Light-weight structures produced by laser beam joining for future applications in automobile and aerospace industry," *Journal of Materials Processing Technology*, vol. 115, no. 1, pp. 2–8, 2001.
[40] A. P. Mouritz, *Introduction to Aerospace Materials*. Elsevier, 2012.
[41] W. M. Kays and A. L. London, "Compact heat exchangers," Krieger Pub Co; Subsequent edition (January 1, 1998) 1984.
[42] A. F. Ardeshir Farshidianfar, *Plate Heat Exchanger*. Nama Publication, 2014.
[43] R. Shah and A. Wanniarachchi, "Plate heat exchanger design theory,", VKI, Industrial Heat Exchangers, p. 37, 1991.
[44] M. Awais and A. A. Bhuiyan, "Heat and mass transfer for compact heat exchanger (CHXs) design: A state-of-the-art review," *International Journal of Heat and Mass Transfer*, vol. 127, pp. 359–380, 2018.
[45] A. F. Anooshiravan Farshidianfar, *Advanced Plate Heat Exchangers Design*. Parto Negar Toos Publication, 2018.
[46] M. M. Abu-Khader, "Plate heat exchangers: Recent advances," *Renewable and Sustainable Energy Reviews*, vol. 16, no. 4, pp. 1883–1891, 2012.

5 Laser Beam Welding
Modeling

5.1 INTRODUCTION

Permanent joining techniques, such as welding, are critical and essential manufacturing methods that can enhance product design and reduce production costs. However, these techniques encounter various challenges. The Laser Beam Welding (LBW) process, a relatively newly developed method, has garnered significant interest in recent decades due to its advantages, including high accuracy, fast weld speed, and localized high concentrated heat input [1,2].

The practical applications of LBW span across numerous industries, such as aerospace, automotive, and shipbuilding, and involve a variety of materials, including steels, titanium alloys, and aluminum alloys. Different techniques, including autogenous welding, laser-arc hybrid welding, and filler wire welding, have been extensively studied within the LBW process [3,4].

Moreover, the LBW process is influenced by various factors, such as welding processing parameters, laser beam quality, interactions between irradiation and the material, and environmental fluctuations. Understanding and predicting the characteristics of the LBW process and their impact on joint performance remain challenging due to the wide distribution of results. Despite these challenges, the advantages of LBW make it a promising method for addressing manufacturing needs and advancing various industries [5,6].

Under extreme laser beam welding (LBW) thermal cycles, the metallurgical characteristics of the weld joint and surrounding internal stress undergo significant changes, distinguishing them from the base metal (BM) [7–9]. Noteworthy challenges associated with the LBW process include the control of various characteristics in welded parts, encompassing geometrical, metallurgical, and mechanical aspects, as well as addressing weld defects [10,11].

Weld geometrical characteristics, including the geometry of the weld or fusion zone (FZ), play a crucial role. Studies indicate that changes in weld geometrical characteristics have a substantial impact on other aspects of the weld. For example, increasing geometrical features such as the radius of weld toes and decreasing flank angle values have been shown to enhance the fatigue life and fatigue strength of butt welds [12]. The fracture peak load of the weld demonstrates an increase with an expansion of the FZ dimensions [13]. Additionally, research highlights the significant influence of notch weld geometry on the fatigue failure of the weld [14]. The width, another geometrical characteristic of the weld, directly affects the fracture behavior of the weld [15]. Given that weld performance and other

DOI: 10.1201/9781003492191-5

characteristics are intricately linked to geometrical weld features, numerous studies in the LBW modeling domain consider this characteristic.

Metallurgical properties are the primary factors governing weld metal characteristics, and these properties are notably influenced by the chemical composition of the weld. Various aspects of the metallurgical characteristics include grain size, alloying elements, and microstructures that arise from the welding process within the weld joints. The crucial role of grain size in determining mechanical properties has been emphasized in research [16]. Smaller grain sizes and the presence of martensite in the weld FZ contribute to higher hardness within that zone. However, the formation of martensite, known for its poor corrosion resistance, can result in diminished corrosion resistance in the weld metal. Conversely, an increased proportion of the austenite microstructure, known for its good resistance, enhances corrosion resistance within the weld FZ [17].

Studies have also indicated that the toughness of weld metals can be improved by increasing the presence of acicular ferrite [18]. Conversely, the presence of grain boundary ferrite can decrease the toughness of the weld metal [19]. Hence, the influence of alloying elements, microstructure, and grain size on weld performance is a critical topic, underscoring the significant impact of metallurgical characteristics on overall weld performance. This highlights the growing demand for comprehensive studies on models describing these metallurgical characteristics. Such studies are essential for understanding the behavior of these characteristics during welding and their subsequent influence on weld performance. By investigating and comprehending these metallurgical characteristics, it becomes possible to optimize welding parameters, select appropriate materials, and enhance the overall performance and durability of welds.

Weld mechanical characteristics, directly influencing the mechanical behavior of welds, are commonly referred to as key determinants of weld performance. These characteristics exert a significant influence on the overall mechanical integrity of the weld. Research findings suggest that inducing lower levels of residual stress can moderately reduce the strength of the joint [20]. Residual stress, considered a weld mechanical characteristic, is generated due to rapid temperature variations during the welding process and plays a crucial role in the life-cycle behavior of structures, particularly in critical connection regions found in offshore and marine applications [21].

Furthermore, studies have demonstrated that an increase in weld hardness correlates with a decrease in yield strength, impact toughness, and tensile strength [22]. This underscores the pivotal role played by mechanical characteristics in the performance of laser beam-welded components, showcasing their significant impact in various ways. Investigating and controlling the mechanical characteristics of welds is essential for optimizing the performance and reliability of welded components. By considering factors such as residual stress, hardness, yield strength, impact toughness, and tensile strength, informed decisions can be made, and strategies can be developed to improve the mechanical properties of welds, leading to enhanced overall performance.

Accurately modeling the laser beam welding (LBW) process is paramount for achieving the desired characteristics in weld joints. Developing a comprehensive model for LBW is essential to effectively implementing and optimizing these welding techniques. However, modeling the LBW process presents inherent challenges due to the rapid cooling rates and high temperature gradients caused by the high energy density of laser beams.

Moreover, the LBW process involves multiple interconnected physical mechanisms, including phase transition, laser-material interaction and absorption, and energy transfer across solid, liquid, gas, and plasma phases [23]. These mechanisms are strongly coupled and must be considered in the models. Hence, it is imperative to develop proper and applicable models capable of capturing the complex nature of the LBW process. Such models not only facilitate the investigation of the LBW process but also contribute to a better understanding of the underlying mechanisms involved.

By employing accurate and robust models, researchers and practitioners can gain insights into the intricate dynamics of the LBW process. These models enable a comprehensive analysis of process parameters, material interactions, and energy transfer, thereby aiding in the optimization of weld quality and process efficiency. Additionally, understanding the LBW process through modeling enhances our knowledge of the fundamental phenomena occurring during welding, ultimately leading to advancements in the field of laser welding technology.

Despite the increasing demand for high-performance welds in various applications, the literature on detailed modeling of the Laser Beam Welding Process (LBW) focusing on geometrical, metallurgical, and mechanical characteristics remains limited. While some studies have attempted to review the LBW process [24–28], a comprehensive study encompassing different types of LBW process modeling has not yet been reported. Notably, interesting features are found in the existing literature. Several studies have explored dissimilar joining of aluminum alloys to steels [24], laser welding of NiTi [26], strength enhancement of carbon fiber-reinforced thermoplastic composites, and metal dissimilar joints [27], as well as the suppression of solidification cracks in laser welding by controlling grain structure and chemical compositions [28]. However, none of these studies specifically focused on the types of LBW process modeling or the characteristics of laser beam-welded parts as the primary aim of the investigation. As a result, previous research has mainly examined general features of welds rather than the modeling or specific characteristics of laser beam-welded parts.

To address this research gap, there is a clear need for comprehensive studies that specifically delve into LBW process modeling and elucidate the geometrical, metallurgical, and mechanical characteristics of laser beam-welded parts. By focusing on these aspects, researchers can develop a deeper understanding of the LBW process and its influence on weld quality and performance. Additionally, such studies will provide valuable insights and guidelines for optimizing the LBW process to achieve desired weld characteristics in diverse application areas.

In conclusion, the understanding and application of laser beam welding process modeling (LBWPM) have demonstrated a positive impact on the performance of laser beam-welded joints. However, there is still a need for further research and comprehensive reviews in the field of LBWPM. In this chapter, our aim is to provide a thorough review of LBWPM, with a focus on various characteristics of weld joints, including geometrical, metallurgical, and mechanical aspects.

5.2 TYPES OF LBW PROCESS MODELING

Laser beam welding (LBW) is a highly sensitive process influenced by various factors that contribute to the creation of diverse joint characteristics. Consequently, the modeling of this process can be categorized into different aspects, covering types of characteristics, monitoring, length scales, and methods. In this section, each of these classifications will be introduced and thoroughly discussed.

The first classification pertains to the types of characteristics crucial for LBW modeling, encompassing geometrical, metallurgical, and mechanical attributes. These attributes play a significant role in determining the quality and performance of welded joints. We will explore how these characteristics are influenced by various factors within the LBW process.

The second classification focuses on monitoring techniques employed during the LBW process. Pre-process, in-process, and post-process monitoring techniques ensure optimal weld quality and detect potential issues or defects. We will discuss the significance of each monitoring stage and the corresponding methods utilized for effective monitoring.

The third classification involves the length scale considered in LBW modeling, encompassing processes operating at different scales, including macroscale, mesoscale, and microscale. We will examine how each length scale affects the welding process and the corresponding modeling approaches used to accurately capture the associated phenomena.

Finally, we will discuss the different methods employed for LBW modeling, broadly classified as empirical-based (EB) or theoretical-based (TB) approaches. Empirical models rely on experimental data and observations, while theoretical models utilize fundamental principles and governing equations to simulate the welding process. We will delve into the advantages and limitations of each modeling method and their relevance in the context of LBW.

5.2.1 Characteristics of the Weld Joint Approach

Given that laser beam welding is a thermal-based welding process, it is crucial to consider various characteristics, namely geometrical, metallurgical, and mechanical, in the process modeling. This characteristics-based classification of laser beam welding process modeling (LBWPM) is depicted in Figure 5.1.

Geometrical characteristics of the weld, such as weld width [29], weld depth [30], and distortion [31,32], are critical considerations in LBW process modeling.

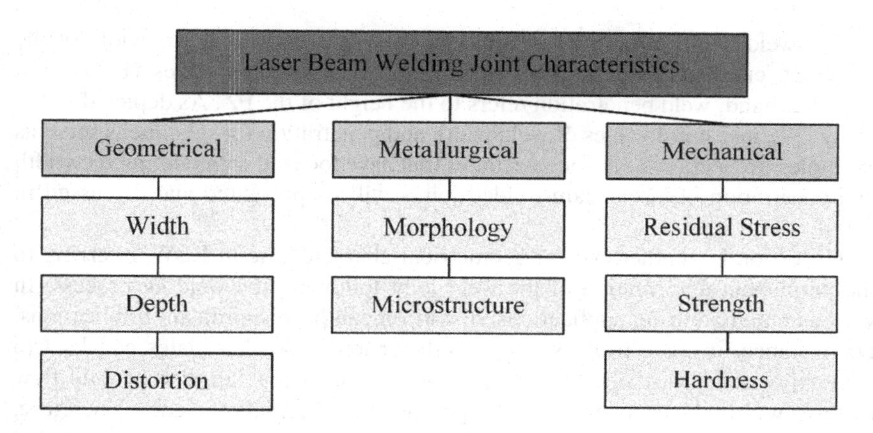

FIGURE 5.1 Types of LBWPM based on weld joint characteristics.

These geometrical characteristics directly impact the physical dimensions and shape of the weld.

The metallurgical characteristics of laser beam-welded parts encompass aspects such as morphology [33] and microstructure [34]. These characteristics describe the arrangement and composition of the materials in the weld joint, which are outcomes of the modeling process.

Mechanical characteristics of the weld, including residual stress [35], strength [36], and hardness [37], play a vital role in determining the performance and functionality of the welded parts. These mechanical properties are considered outputs of the modeling process, and they directly influence the structural integrity and durability of the weld joint.

In the following subsections, we will delve into each of these types of modeling, providing detailed discussions and analyses. By considering these characteristics in LBWPM, researchers and practitioners can gain a comprehensive understanding of the weld quality and performance, enabling them to optimize the welding process for desired outcomes.

5.2.1.1 Geometrical Characteristics

Geometrical characteristics of the weld provide valuable insights into the welding process and are often the focus of fundamental Laser Beam Welding Process Modeling (LBWPM) research. Achieving specific geometrical features is crucial for optimizing welding production efficiency. For instance, attaining full penetration and a desirable weld width in a single pass can enhance productivity [38]. Therefore, it is of utmost importance to carefully consider and control the geometrical aspects of the welding process.

It is worth noting that weld geometry varies across the entire thickness, and assessing the modeling based solely on penetration and surface width can be an oversimplification. To capture the complexity of the weld bead, it is necessary to investigate various geometrical features. However, three key features stand out as the most significant: weld width (W), weld penetration (P), and distortion (D).

The weld width in laser beam welding (LBW) is defined as the width of the FZ (FZ), encompassing both the fusion and heat-affected zones (HAZ). On the other hand, weld penetration refers to the height of the FZ. As depicted in the study [39], measured values of weld width and penetration for specimens serve as examples. In Section 5.3.1, recent studies that have focused on modeling the width and penetration of laser beam-welded joints will be presented and discussed in detail.

Distortion is another critical geometrical characteristic in LBW, referring to the permanent deformation of the weld joint following the welding process. In various manufacturing applications, distortion can have significant implications. For instance, in plate heat exchangers, distortion in welded plates can lead to assembly issues. Furthermore, the presence of distortion can disrupt fluid flow patterns within the exchanger, resulting in reduced thermal efficiency. Therefore, minimizing distortion in welded plates is crucial. Studies have classified distortion of weld joints into two general categories: in-plane (transverse contraction, longitudinal contraction, and rotational distortion) and out-of-plane (angular distortion, longitudinal bending, and buckling distortion). These categories are illustrated schematically in the study [40]. In Section 5.3.2, notable studies that have modeled the deformation in LBW are examined and compared in various aspects.

5.2.1.2 Metallurgical Characteristics

Weld metallurgical characteristics play a crucial role in determining the operational performance of welding joints and serve as fundamental evaluation criteria for joint characteristics. The metallurgical characteristics of the FZ in laser beam welding (LBW) significantly impact other characteristics, particularly the mechanical properties of the weld. It has been observed that a coarse columnar structure in the FZ adversely affects the mechanical characteristics of the weld [41]. Conversely, the presence of finely equiaxed grains in the FZ offers two notable advantages: first, it reduces the susceptibility to solidification cracking during the LBW process, and second, it enhances the mechanical properties [42]. Furthermore, the microstructure, which directly relates to the mechanical performance of welds, is predominantly influenced by factors such as metal compositions, cooling rates, and temperature gradients [43].

Metallurgical characteristics in LBW can be categorized into two main groups:

1. Weld Metal Solidification (WMS): This includes the microstructure within grains (MWG) or solidification mode (SM) and grain structure size (GSS) or morphology.
2. Post-Solidification (PS): This category considers phase transformations (PT), which will be discussed in more detail in the following subsections. The classification of metallurgical characteristics provides a comprehensive understanding of the LBW process and its impact on the resulting weld properties.

5.2.1.3 Solidification Mode and Phase Transformations

During the solidification process of materials, different morphologies can be observed, including planar, cellular, columnar dendritic, and equiaxed dendritic structures. These morphologies are considered in studies [44–45]. A study [42] has identified a relationship that needs to be satisfied to achieve a stable solidification mode under controlled conditions. The relationship can be expressed as follows:

$$\frac{G}{R} \geq \frac{\Delta T}{D_L} \tag{5.1}$$

The solidification process is influenced by several factors, including the temperature difference (ΔT) along the boundary layer and the liquid diffusivity coefficient (D_L). The temperature difference ΔT is defined as the difference between the liquidus temperature (T_L) and the solidus temperature (T_S). Additionally, the movement speed of the solid-liquid interface within the middle layer is referred to as the "Solidification growth rate" (R).

The "Temperature gradient" (G) in the liquid metal represents the difference in the temperature profile. As the solidification growth rate and temperature gradient increase, the solidification modes undergo transitions from planar to cellular, cellular to columnar, and finally from columnar to equiaxed dendrite structures. The relationship between these parameters plays a crucial role in determining the solidification behavior and resulting microstructure during the welding process.

The laser beam welding (LBW) process, being a high-energy laser material processing technique, is characterized by rapid cooling and high growth rates, coupled with a small temperature gradient. As a result, the solidification process in LBW predominantly leads to the formation of columnar or equiaxed dendritic structures [46]. In the case of stainless steel, which is widely used in LBW, the solidification modes and resulting microstructures are influenced by the Cr_{eq}/Ni_{eq} ratio for the steel material. This ratio is defined based on various studies and is presented in Table 5.1. Additionally, the study [47] provides further insights into the relationship between Creq/Nieq and solidification modes and microstructure in LBW for stainless steel. These findings highlight the significance of the Creq/Nieq ratio in determining the solidification behavior and resulting microstructure, contributing to the understanding and optimization of the LBW process for stainless steel.

TABLE 5.1
The Relationship of Cr_{eq}/Ni_{eq}

Row	Formula	References
1	$(Cr+1.37Mo+1.5Si+2Nb+3Ti)/(Ni+0.31Mn+22C+14.2N+Cu)$	[48]
2	$(Cr+Mo+1.5Si+0.5Nb)/(Ni+0.5Mn+30(C+N))$	[49]
3	$(Cr+Mo+0.7Nb)/(Ni+35C+20N+0.25Cu)$	[50]

5.2.1.4 Morphology

Grain size and morphology are essential metallurgical characteristics of the FZ in the laser beam welding (LBW) process. It has been observed that increasing the product of the temperature gradient (*G*) and the solidification growth rate (*R*), referred to as the "Cooling Rate" (*CR*), leads to a decrease in grain size. Research studies have shown that higher cooling rates result in finer cellular or dendritic structures, indicating a refined grain size [42].

In LBW process modeling, three main parameters are considered to characterize the morphology or grain size: the eutectic spacing (λ_E), primary dendritic arm spacing or λ_1, and secondary dendritic arm spacing or λ_2. These parameters play a significant role in determining the morphology and grain size in the LBW process [51].

5.2.1.3 Mechanical Characteristics

The strength of a welded joint is critical, as it should be at least as strong as the weakest of the two metals being joined to prevent failure during welding [52]. Therefore, investigating the mechanical characteristics of welded joints is crucial to determining their performance and functional properties. Welded joints in sensitive and high-value applications, such as power steam stations, chemical tankers, and chemical plants, require appropriate mechanical characteristics, including tensile strength, hardness, and residual stress [53].

Tensile strength is an important parameter used to assess the tensile properties of weldments [54]. The tensile strength of a weld joint can be influenced by factors such as loading capacity and weld quality. Poor loading capacity and weld quality result in lower tensile strength, while higher toughness in the weld metal can lead to improved tensile strength [55]. Section 5.5.1 will present studies focused on modeling weld joint strength, including ultimate tensile strength (UTS), yield stress, and the strain corresponding to UTS.

Maintaining suitable hardness levels in welded joints is crucial [56]. Various investigations have shown changes in weld joint hardness as a result of the laser beam welding process. The LBW process can significantly increase the hardness of the weld metal and the coarse prior austenite grain bainite region compared to the BM hardness. However, post-weld heat treatment variants can reduce the hardness of these regions in the welded joints to the level of the BM [54]. It has been reported that the hardness is approximately 60% higher in the FZ compared to the BM, with the maximum hardness observed in the supercritical HAZ due to its refined microstructure. No HAZ softening was observed [57]. Section 5.5.2 will present models of hardness profiles in LBWPM.

Residual stress resulting from welded joints is another important factor to consider in the mechanical characteristics of laser beam-welded structures. Tensile residual stresses in welded structures can reach levels as high as the material's yield strength and have a detrimental effect on weld performance. The combination of tensile welding residual stresses and operational stresses can promote fatigue failure. Conversely, compressive residual stresses can have a beneficial effect on fatigue life. However, spectrum loading may relax part of the residual stress field, which can affect the final fatigue life [58]. Section 5.5.3 will discuss in-depth research on modeling residual stress resulting from the LBW process.

5.2.2 Monitoring Approach

In this section, the categorization of laser beam welding process modeling (LBWPM) is organized around the monitoring approach. The primary objective of monitoring the LBW process is to detect defects and ensure the integrity of the weld joint. The monitoring process in LBWPM is classified into three distinct stages: (1) pre-process, (2) in-process, and (3) post-process [59]. These stages are visually represented in the study [59], and a comprehensive summary of LBWPM, encompassing monitoring aims, signals, and technologies at different stages, is outlined in the study [59].

The pre-process models aim to capture specific weld features before applying the laser beam to the workpiece. Utilizing optical signals with machine vision and laser triangular technologies, these models address weld geometrical characteristics such as seam tracking and scanning the joint gap between workpieces, ensuring accurate joint formation [60].

In-process models focus on characteristics generated during the LBW process in the weld zone. Employing various technologies, these models use acoustical, optical, electrical, or thermal signals to detect weld geometrical characteristics (e.g., weld width, penetration, and keyhole geometry) and defects (e.g., surface cracking) in real-time. Analyzing these signals and characteristics enables the prediction and adjustment of weld seam quality [61].

Post-process models encapsulate characteristics obtained after completing the LBW process, encompassing geometrical aspects (e.g., weld width, penetration, and distortion), metallurgical properties (e.g., microstructure, grain size), mechanical properties (e.g., yield strength, fracture force), and defects (e.g., internal cracking, lack of fusion, porosity) [62]. Employing various technologies such as machine vision, destructive and nondestructive inspection, metallurgical tests, and laser triangulation, supported by acoustical and optical signals, these models are detailed in the study [59].

It is crucial to emphasize that creating an effective LBW process model requires monitoring and collecting data using one of the mentioned technologies. Subsequently, the data undergoes analysis using different methods and models. Thus, LBW process modeling can be classified based on monitoring methods into three categories: pre-process, in-process, and post-process, as depicted in Figure 5.2. Furthermore, these models can be analyzed in two categories: (1) static and (2) dynamic (offline or online). Static models are time-independent, while dynamic models are time-dependent, with the output characteristics of static models remaining unaffected by time.

Dynamic models in LBW process modeling can be further categorized into two types: offline and online models, both relying on time-dependent

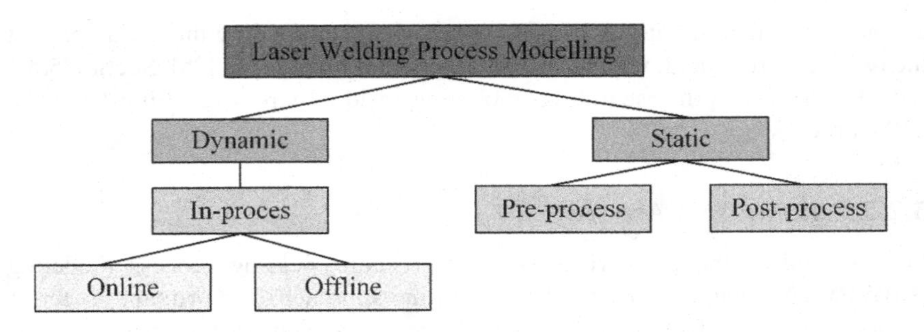

FIGURE 5.2 Classification of Laser Beam Welding Modeling studies based on process monitoring stage.

characteristics. Online models consider signals throughout the entire LBW process, providing real-time monitoring and adapting to fluctuations, ensuring a comprehensive understanding of momentary environmental changes impacting the output characteristics. In contrast, offline LBWPM involves detecting signals only from the initial step or specific selected steps and incorporating them into the model. Characteristics are then provided based on these selected steps, excluding real-time fluctuations throughout the process. Consequently, offline models do not capture the dynamic variations occurring in real-time and are limited to the information available from the chosen steps.

5.2.3 Length Scale

Despite the widespread use of laser beam welding (LBW) and the considerable research devoted to LBW modeling, the fundamental physics of this process remains an area of ongoing investigation. This is attributed to the intrinsic complexity of the LBW process, where numerous coupled physical phenomena take place within a confined region of the melt pools. Process disturbances can introduce variations in the weld zone, leading to non-equilibrium physical and chemical metallurgical processes, thereby compounding the challenges faced in understanding the process.

For in-depth exploration, researchers have identified five distinct mechanisms associated with LBW, as detailed in the study [63]: (1) absorption, (2) heat conduction, (3) vapor dynamics, (4) melt dynamics, and (5) phase transitions. Each of these mechanisms contributes to shaping the overall behavior of the LBW process, adding to its complexity and necessitating comprehensive investigations. Through the categorization and study of these physical phenomena, researchers strive to enhance their understanding of the underlying mechanisms and develop more accurate models for LBW. This ongoing research aims to address the challenges posed by the intricate interplay of these coupled phenomena, ultimately improving the overall performance and reliability of the LBW process.

Study [63] delves into the absorption mechanism within the LBW context, covering Frensel absorption, multiple reflections, vapor and plasma absorption, and temperature-dependent optical properties. The heat conduction mechanism in LBW involves convective and conductive heat flux as well as melting and evaporation enthalpies. The vapor dynamic mechanism includes two notable phenomena: pressure waves and the Bernoulli effect. The phase transition mechanism, one of the most intricate aspects of LBW, encompasses melting and solidification, evaporation, and condensation. The melt dynamics mechanism in LBW includes melt expulsion, spilling formation, Marangoni convection, and temperature-dependent material properties, as discussed in Section 5.2.1.2.

The fluid dynamics of the melt pool during LBW, as illustrated in the study [64], present both vertical and horizontal perspectives. The velocity field demonstrates liquid melt waves cascading down the front of the keyhole, leading to periodic variations in keyhole diameter and resulting in keyhole oscillations. The liquid melt accelerates around the keyhole, encountering backflow from the rear of the melt pool at approximately two-thirds of its length, resulting in turbulence in the lower rear region. The flow pattern of the melt pool in the upper part exhibits a more laminar behavior. Additionally, the study [63] provides insights into ion, vapor pressure at the interface, and mass flux between phases.

To streamline the LBW process, certain phenomena are occasionally decoupled or overlooked in some models [63]. Many models concentrate on analyzing only one of the geometrical, mechanical, or metallurgical characteristics or physical phenomena of LBW. For instance, when laser welding magnesium alloys, the conduction, radiation, and convection mechanisms are considered to model geometrical characteristics like weld width and depth, as outlined in the study [65]. Another study introduces an integrated process-structure-properties-performance model that incorporates the metallurgical characteristics of the weld, considering Marangoni flow, as highlighted in the study [66].

Upon comparison of parts of the study [65,66], it becomes evident that LBW process modeling can be categorized based on the length scale into three groups: (1) macroscale, (2) mesoscale, and (3) microscale, as illustrated in the study [67]. Macroscale modeling, with a length scale of approximately $100–10^{-2}$m, involves macro-thermomechanics, process modeling, and performance modeling. Mesoscale models operate within the scale of approximately 10^{-2} to 10^{-4}m and encompass thermo-fluid dynamics and meso-mechanics mechanisms. Microscale models, with a scale of approximately 10^{-4} to 10^{-6}m, focus on microstructure modeling. Consequently, it can be inferred that each mechanism and characteristic modeling is categorized within different length scale models.

5.2.4 Methods Approach

Classified by the approach employed for modeling and solving the laser beam welding (LBW) process, two primary types emerge: (1) EB and (2) TB, as visually depicted in Figure 5.3. Theoretical-based LBW models are constructed upon the foundational theory governing the LBW process and can be further divided into two categories: (1) exact models and (2) numerical models.

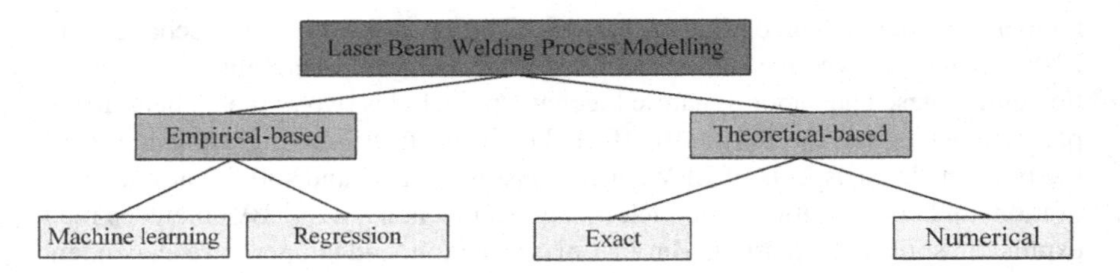

FIGURE 5.3 Classification of LBWPM based on methods approach.

Exact models aim to represent the process outputs by considering the equilibrium between the input, output, and waste energy of the LBW process and solving the resulting equations using precise analytical methods. Numerical models, such as finite element methods, numerical software, and other techniques, are also utilized in certain studies to solve the theoretical equations governing the process. Section 5.2.4.3 presents some of the governing equations that can be considered in LBW theoretical-based models.

The second category is EB LBW process models, broadly divided into two groups: (1) machine learning or modern approaches and (2) regression or traditional approaches. Machine learning methods, such as fuzzy logic, neural networks, genetic algorithms, etc., categorize models falling under the machine learning group when employed to model the LBW process. Conversely, the regression model, a simpler method, involves deriving a model by fitting a curve between the input and output variables of the LBW process.

5.2.4.1 Some of the Applicable Mathematical Equations

To construct theoretical-based models for the laser beam welding (LBW) process, this section explores several governing equations. Key equations include the conservation equations of continuity, momentum, energy, and mass fraction, each playing a vital role in elucidating the fundamental principles at play during laser beam welding. Refer to equations (5.2–5.5) for an outline of these equations, which ensure the conservation of physical quantities throughout the welding process. Notably, the LBW process encompasses various heat transfer mechanisms, such as conduction, convection, and radiation, extensively discussed in Section 5.2.3. These mechanisms are crucial for understanding the distribution of heat and energy within the welding zone.

Additionally, forces acting on the workpiece during the LBW process merit consideration. Surface tension, buoyancy force, Marangoni convection, and gravity represent significant forces that contribute to the overall dynamics and behavior of the molten material. The interplay of these forces significantly influences the quality and characteristics of the weld. Therefore, a comprehensive understanding and accurate incorporation of these forces are imperative for precise and effective modeling of the LBW process [68]. Some applicable equations are provided in the following equations (5.2)–(5.5): continuity equation, momentum equation, energy equation, and mass fraction equation, respectively [69].

$$\frac{\partial \rho}{\partial t}+\frac{\partial}{\partial x}(\rho v_x)+\frac{\partial}{\partial y}(\rho v_y)+\frac{\partial}{\partial z}(\rho v_z)=0 \tag{5.2}$$

$$\begin{cases} \frac{\partial}{\partial t}(\rho v_x)+\frac{\partial}{\partial x}(\rho v_x^2)+\frac{\partial}{\partial y}(\rho v_x v_y)+\frac{\partial}{\partial z}(\rho v_x v_z) \\ =\frac{\partial}{\partial x}\left(\mu\frac{\partial v_x}{\partial x}-p\right)+\frac{\partial}{\partial y}\left(\mu\frac{\partial v_x}{\partial y}\right)+\frac{\partial}{\partial z}\left(\mu\frac{\partial v_x}{\partial z}\right)-M\frac{(1-f_L)^2}{f_L^3+B}v_x \\ \frac{\partial}{\partial t}(\rho v_y)+\frac{\partial}{\partial x}(\rho v_x v_y)+\frac{\partial}{\partial y}(\rho v_y^2)+\frac{\partial}{\partial z}(\rho v_y v_z) \\ =\frac{\partial}{\partial x}\left(\mu\frac{\partial v_y}{\partial x}\right)+\frac{\partial}{\partial y}\left(\mu\frac{\partial v_y}{\partial y}-p\right)+\frac{\partial}{\partial z}\left(\mu\frac{\partial v_y}{\partial z}\right)-M\frac{(1-f_L)^2}{f_L^3+B}v_y \\ \frac{\partial}{\partial t}(\rho v_z)+\frac{\partial}{\partial x}(\rho v_x v_z)+\frac{\partial}{\partial y}(\rho v_y v_z)+\frac{\partial}{\partial z}(\rho v_z^2) \\ =\frac{\partial}{\partial x}\left(\mu\frac{\partial v_z}{\partial x}\right)+\frac{\partial}{\partial y}\left(\mu\frac{\partial v_z}{\partial y}\right)+\frac{\partial}{\partial z}\left(\mu\frac{\partial v_z}{\partial z}-p\right)-M\frac{(1-f_L)^2}{f_L^3+B}v_y \\ +\rho g(T-T_0) \end{cases} \tag{5.3}$$

$$\begin{aligned} &\frac{\partial}{\partial t}(\rho H)+\frac{\partial}{\partial x}(\rho v_x H)+\frac{\partial}{\partial y}(\rho v_y H)+\frac{\partial}{\partial z}(\rho v_z H) \\ &=\frac{\partial}{\partial x}\left(K\frac{\partial T}{\partial x}\right)+\frac{\partial}{\partial y}\left(K\frac{\partial T}{\partial y}\right)+\frac{\partial}{\partial z}\left(K\frac{\partial T}{\partial z}\right)+q_{laser} \end{aligned} \tag{5.4}$$

$$\begin{aligned} &\frac{\partial}{\partial t}(\rho w)+\frac{\partial}{\partial x}(\rho v_x w)+\frac{\partial}{\partial y}(\rho v_y w)+\frac{\partial}{\partial z}(\rho v_z w) \\ &=\frac{\partial}{\partial x}\left(\rho D\frac{\partial w}{\partial x}\right)+\frac{\partial}{\partial y}\left(\rho D\frac{\partial w}{\partial y}\right)+\frac{\partial}{\partial z}\left(\rho D\frac{\partial w}{\partial z}\right) \end{aligned} \tag{5.5}$$

The nonlinear heat transfer governing equation, which can be considered in conducting the LBW process modeling, is as follows [70]:

$$\rho c\frac{\partial T}{\partial t}(x,y,z,t)=-\nabla.q(x,y,z,t)+Q(x,y,z,t) \tag{5.6}$$

where ρ is the density of the materials, c is the specific heat capacity (J/(g°C)), T is the current temperature (°C), q is the heat flux vector (W/mm^2), Q is the internal heat generation rate (W/mm^3), x, y, and z are the coordinates in the reference system (mm), t is the time (s), and ∇ is the spatial gradient operator. The nonlinear isotropic Fourier heat flux constitutive equation is employed [70]:

$$q = -k\nabla T \tag{5.7}$$

where k is the temperature-dependent thermal conductivity (J/(mm s °C)).

5.2.4.2 Physical Material Properties Modeling

In addition to the detailed mechanisms discussed in Section 5.2.3, the laser beam welding (LBW) process intricately involves a solidification process that seamlessly blends thermodynamic principles and kinetic reactions. This intricate interplay results in the nuanced formation of various metallurgical, mechanical, and geometrical characteristics within the weld zone [46,71]. The schematic diagram illustrating this deductive process is presented in study [70]. Subsequently, equilibrium phases, such as ferrite and austenite in metallurgical characteristics, are precisely determined utilizing thermodynamic process properties like melt pool temperature and phase diagrams. Venturing into the realm of kinetic reactions, the dynamic evolution of phase changes and morphology (i.e., metallurgical characteristics) during the thermal LBW process is meticulously captured. This is accomplished by considering kinetic reaction parameters such as cooling rate and heating rate and utilizing indispensable tools like time-temperature transfer diagrams and solidification diagrams, as vividly depicted in the study [70].

Furthermore, detailed insights into metallurgical and crystallographic characteristics, such as atomic structure and material arrangement, can be gleaned through scanning electron microscope tests [71]. The knowledge derived from crystallographic details and metallurgical characteristics is instrumental in determining the mechanical and geometrical properties of the weld, acting as key indicators of the LBW process performance. Consequently, an in-depth understanding of the temperature specifics of the process is crucial for a comprehensive evaluation of the weld joint's performance.

As depicted in the study [71], the modeling of the laser beam welding process adopts a thermo-mechanical-metallurgical approach that pivots on temperature dynamics. Therefore, the consideration of temperature-dependent physical material properties becomes imperative in the modeling process. These properties neatly fall into two categories: (1) thermal properties and (2) mechanical properties. Thermal material properties encompass factors such as thermal conductivity, density, and specific heat, while mechanical material properties include parameters like Young's modulus, yield strength, Poisson's ratio, and thermal expansion coefficient. Study [70] visually presents the thermal and mechanical properties, respectively, using low-carbon steel (Q235) as an illustrative example. Some studies [72–74] have advocated for the utilization of JMatPro software to ascertain material properties in LBWPM methods.

5.3 GEOMETRICAL CHARACTERISTICS MODELING CASE STUDIES

As emphasized in Section 5.2.1.1, addressing the modeling of geometrical characteristics is pivotal in understanding the intricacies of the laser beam welding (LBW) process. This section will delve into recent case studies that specifically concentrate on modeling distinct geometrical features, namely weld width, penetration, and distortion. Each of these characteristics will be thoroughly examined in separate subsections, providing a comprehensive overview of the advancements and challenges associated with their modeling.

5.3.1 Weld Width and Penetration

Weld width and penetration are pivotal geometrical characteristics extensively studied in the realm of laser beam welding (LBW) modeling. LBW is classified into five penetration modes, denoted as P modes, based on the shape of weld penetration. These modes include:

1. Conduction
2. Transition
3. Keyhole: Partial
4. Keyhole: Full (Wide Root)
5. Keyhole: Full (Thin Root)

Table 5.2 presents the weld shapes and schematics corresponding to each of these penetration modes. The exploration and understanding of these modes contribute significantly to advancing LBW modeling studies.

The conduction mode in LBW offers advantages such as the absence of cracks, porosity, undercut, and spatter [90]. However, the keyhole laser beam welding mode is more versatile and finds broader applications compared to the conduction mode [91,92]. These modes are distinguished based on the applied power density on the weld area, calculated as the laser power divided by the laser beam area.

In the keyhole mode, the power density is sufficiently high to vaporize material, creating a hole in the melt pool. In contrast, the conduction mode operates with a lower power density, preventing vaporization [93]. The boundary between the keyhole and conduction modes is not well-defined and is characterized as a transition mode. This transition is influenced by material thermal properties, including melting temperatures, thermal conductivities, density, specific heat capacity, latent heat of melting, and latent heat of vaporization [76].

The study [76] delves into the conduction, transition, and keyhole modes based on power density for various materials, including stainless steel 304L, aluminum 2024-T3, and mild steel S355. Understanding and characterizing these modes is critical for optimizing the LBW process for different materials and applications.

The schematics, both experimental and numerical, depict three keyhole laser beam welding modes: partial penetration, full penetration with a wide root, and

TABLE 5.2
Classification of the Laser Beam Welding Process Based on Penetration Modes (P Modes)

Row	P Mode	Weld Shape	Schematic	References
1	Conduction	Horizontal elliptic: Wide at the top+shallow penetration and very low aspect ratio		[75,76]
2	Transition	Vertical elliptic: Wide at the top+medium penetration and low aspect ratio	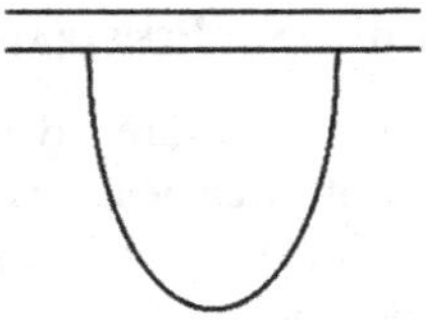	[75,76]
3	Keyhole: Partial	Bell shape: Wide at the top+ Semi-cylindrical in penetration	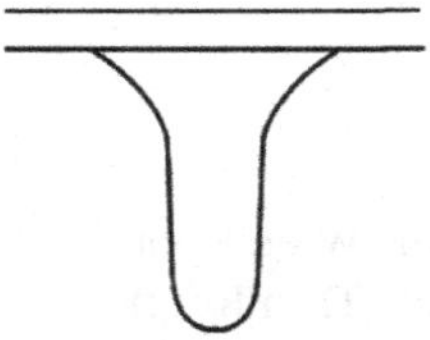	[77–81]
4	Keyhole: Full (Wide Root)	Hourglass shape: Wide at the top and root+ Semi-cylindrical in penetration	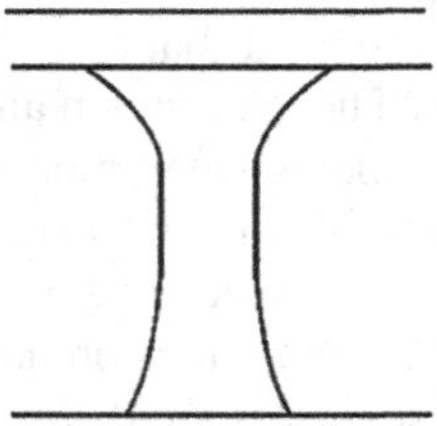	[77,82–86]
5	Keyhole: Full (Thin Root)	Wide at the top+Semi-cylindrical in penetration+Minor widening at the root	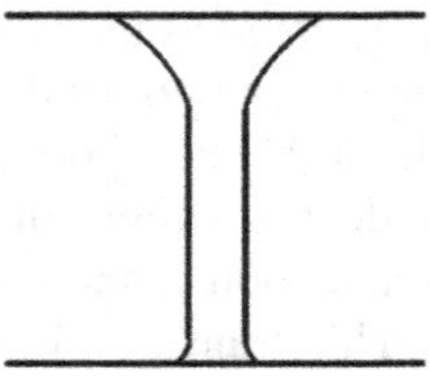	[77,83,85,87–89]

full penetration with a thin root, as illustrated in the study [77]. In this study, conductive melting is achieved by redirecting the downward flow upwards, resulting in partial penetration. For full penetration with a wide root, the upward flow is sufficient to prevent melt from flowing backward, and the melt solidification behind the keyhole is not intense enough to form a root concavity, resulting in a flat root. A wider melt width allows more melt to flow after the keyhole exit.

This excess melt flow can lead to inappropriate amounts of melt reaching the end of the melt pool, causing the formation of humps. Understanding these modes is essential for optimizing welding parameters and achieving desired weld characteristics [77].

The investigation of various modeling approaches for five laser beam welding penetration modes and weld widths has been a focus of research over the past decades. A comprehensive summary of these studies, organized by the BM material, thickness, modeling types (monitoring, approach, and methods), weld width, weld penetration, and model verification error, is presented in Table 5.3. In this table, the abbreviations Post, Off, Num, ML, Reg, W, and P correspond to post-process, offline, numerical, machine learning, regression, weld width, and weld penetration of the LBW process, respectively. This tabulated information provides a comprehensive overview of the diverse modeling efforts in the context of laser beam welding.

The study on melt pool behavior and transmission mechanisms during laser beam welding (LBW) of dissimilar metals employed a 3D transient numerical model (FLUENT). This model incorporated fluid flow, heat transfer, keyhole evolution, and mass transport aspects [94]. The investigation considered the influences of recoil pressure and surface tension on the keyhole wall, as well as factors like convection, diffusion, and keyhole formation on mass transfer. Various processing parameters were taken into account, including laser power (1,800–2,000 W), welding speed (0.055–0.075 mm/s), focused beam, and heat input (26.67–32.73 J/mm). Model validation was performed against another study with a heat input of 36.3 J/mm [95]. For additional simulation details, refer to the study [94].

The study reported maximum geometrical characteristics, with weld width and weld penetration recorded at 1.383 and 1.4 mm, respectively. The molten pool exhibited a maximum liquid velocity of 30.1 m/s and a temperature of 3,800 K under a laser power of 2,000 W. The findings suggest that convective heat transfer played a dominant role, with convection and diffusion identified as the primary mechanisms of metal mass transport. The study concludes that a reduction in heat input per unit area leads to decreased fluid flow, element diffusion, and thickness of the intermetallic transition layer. Another numerical model, utilizing ANSYS-FLUENT software in conjunction with a high-speed camera, investigated the geometrical characteristics of the weld and its correlation with inconsistent thermodynamic behaviors of the keyhole [96]. Further details can be found in Table 5.3.

A study employing computational fluid dynamics (CFD) with ANSYS software, utilizing a numerical model with a hybrid conical-cylindrical heat source (as presented in the study [97]), delved into the heat transfer, molten fluid flow, weld pool dynamics, and cooling rate phenomena [97]. Stainless steel 316LN with a thickness of 5.5 mm served as the material, while processing parameters involved laser power ranging from 1 to 3.5 kW, resulting in variations of Maragoni number (1,813–22,623) and Peclect number (26.38–135.08). Accounting for heat loss, including convection and radiation heat transfer, the maximum error between experimentally measured and predicted model outcomes was 11%.

TABLE 5.3
Summary of Presented Studies Based on Weld Width, Penetration and Types of Modeling and Error

				Types of Modeling Based on			Max			
Row	Year	Material	Thickness (mm)	Monitoring	Approach	Method	W (mm)	P (mm)	Error (%)	References
1	1996	Mild steel 1010-1035	NR	Post	Exact	2D heat conduction	2.8	6.5	NR	[100]
2	1997	Austenitic stainless steel	NR	Off	Exact	Energy absorption mechanism	5	5	37	[101]
3	1999	SS 1020	9.5	Off	Num	FDM	2	9.1	NR	[102]
4	2000	Steel 1018	3.17	Off	Num	FEM	3.8	NR	NR	[103]
5	2001	Al 5182	1	Off	Exact	Heat conduction	0.44	1.35	5.4	[104]
6	2002	Steel	1.6–2	On	ML	Fuzzy	1.77	NR	NR	[105]
7	2004	Austenitic stainless steel	6.5	Post	ML	Fuzzy logic/ANN	NR	6.5	0.45	[106]
8	2004	TC1	1.5	Off	Num	FEM	0.6	1.5	NR	[107]
9	2005	Medium Carbon Steel	5	Post	Reg	ANOVA	2.703	4.407	6.11	[108]
10	2008	WE43	10	Off	Num	FLUENT	2.5	10	NR	[65]
11	2010	SS 304	1.6	Post	ML	ANN	1.45	0.96	4.6	[109]
12	2010	SS 304	6	Post	Reg	ANOVA	NR	3	NR	[110]
13	2010	AA5083	5	Off	Num	ANSYS	2.8	NR	10	[111]
14	2011	SS 304	6	Post	Reg	ANOVA	NR	3.4	10	[112]
15	2015	SS 304/ Mg AZ31B	2	Post	ML	ANN	NR	2.1	3	[113]
16	2015	NR	NR	Off	ML	GABP	0.77	2.176	5.3	[114]
17	2017	NR	NR	Post	Reg	SVR	3.5	NR	1	[115]
					ML	ANN	3.5		2	

(Continued)

TABLE 5.3 (*Continued*)
Summary of Presented Studies Based on Weld Width, Penetration and Types of Modeling and Error

				Types of Modeling Based on			Max			
Row	Year	Material	Thickness (mm)	Monitoring	Approach	Method	W (mm)	P (mm)	Error (%)	References
18	2018	316L	7	Off	Num	CFD	4.4	7	32.67	[73]
19	2018	Steel S355J2G3	15	Off	Num	CFD/COMSOL	1.5	15	NR	[116]
20	2018	Steel EH36	4	Off	Num	ANSYS	2.125	4	15.6	[117]
21	2019	Ti6Al4V	1	On	ML	PCA/GA	2.12	NR	4.5	[118]
22	2020	TRB	2	On	ML	CNN	NR	1.8	5.4	[51]
23	2020	SS316L	2	Post	ML	GA/BPNN	2.06	NR	0.001	[119]
24	2020	Ti6Al4V	1–1.5	Off	Num	NR	1.5	1	66	[120]
25	2021	HT780 steel	20	Off	Num	FEM	NR	20	NR	[77]
26	2021	Al-Si coated	1.5	Off	Num	CFD/ COMSOL	1.6	1.5	NR	[121]
27	2021	Ti-6Al-4V & Inconel 718	3	Off	Num	NR	1.5	3	22.00	[69]
28	2021	Ti-Grade 5	2	Post	Reg	ANOVA	0.97	0.41	5.00	[122]
29	2021	AA2024	2	Off	Num	COMSOL	2.2	2	5.63	[68]
30	2021	SS316L	10	Off	Num	FLUENT-ANSYS	NR	7.28	NR	[96]
31	2021	DP600-AA6082	3	Off	Num	FLUENT	1.4	1.383	NR	[94]
32	2021	SS316L	1	Off	Reg	ANOVA	1.98	1.51	54.37	[123]
33	2022	SS316NL	5.5	Off	Num	ANSYS	2.2	5.5	11	[97]
34	2022	SS301L	0.6	Off	Num	FE	0.7	0.6	9.41	[124]
35	2022	Al-Si coated	1.5	Off	Num	ABAQUS	NR	1.5	22.00	[125]
36	2022	SS316L	10	Off	Num	SYSWELD	2.98	2.168	5	[126]
37	2022	iPP	2	Off/Post	Num/Reg	COMSOL/ANOVA	1.7	NR	4.42	[127]

(*Continued*)

TABLE 5.3 (*Continued*)
Summary of Presented Studies Based on Weld Width, Penetration and Types of Modeling and Error

				Types of Modeling Based on			Max			
Row	Year	Material	Thickness (mm)	Monitoring	Approach	Method	*W* (mm)	*P* (mm)	Error (%)	References
38	2023	Q235B	7.6	Off	Num	FLUENT/ANSYS	6.95	6.71	14.4	[128]
39	2023	AA5754-AA6005	3	Off	Num	COMSOL	4.1	3	10.1	[33]
40	2023	Al 2219	9	Off	Num	ANSYS	6.9	6.5	5	[129]
41	2023	Al 6082	6	Off	Num	ANSYS-FLUENT	3.20	4.5	16	[130]
42	2023	Invar alloy	3	Off	Num	FLUENT	3.29	3	NR	[131]
43	2023	AA6061	5	Off	Num	NR	3.94	5	12	[132]
44	2023	Al-Cu	0.2 Al/1 Cu	Off	Num	Flow-3D	0.32	0.19	15.6	[133]
45	2023	SS 301L	4	Off	Num	CFD	2.80	3	9.90	[134]
46	2023	S A710	6	Off	Num	FEM	4.32	3.15	0.033	[135]

NR, Not reported; SVR, Support vector regression.

Recoil pressure contour, velocity field, and cross-sectional views of the laser keyhole revealed that once the material surpassed the evaporation temperature, metallic vapor expelled from the cavity formed due to intense laser power density. The vapor flow direction was toward the outflow boundary, and as the keyhole formed, recoil pressure in contact with the vapor plume created a circulation loop along the keyhole wall. Velocity values around the keyhole varied based on processing parameters. The reported maximum pressure for the maximum weld temperature was approximately 133 kPa, aligning with previous studies [98,99].

Analyzing Table 5.3 provides a comprehensive overview of models focused on weld width, penetration, modeling types, and associated errors. It becomes evident that post-process (Post) and in-process, encompassing online (On) and offline (Off) methodologies, have been extensively studied across diverse materials such as various steel types, aluminum alloys, copper alloys, and polymers, with thicknesses ranging from 0.2 to 20 mm.

Different modeling approaches have been employed, including exact methods, numerical techniques like finite element method (FEM)-based software (FLUENT, ANSYS, CFD, and COMSOL), and finite difference method-based models. Additionally, machine learning (ML) techniques such as fuzzy logic, CNN, ANN, GABP, PCA/GA, and EB methods like regression, support vector regression, and ANOVA have been utilized to predict weld width and penetration. The associated error range varies widely, spanning from 0.001% to 66%, reflecting a broad spectrum of model accuracies. In comparing online models, characterized by errors less than 6%, with offline models, it is evident that online models tend to exhibit higher accuracy. This heightened accuracy can be attributed to their consideration of instantaneous changes in the welding process, leading to results that align closely with experimental outcomes. For further details on specific studies, please refer to the respective references provided in Table 5.3.

5.3.2 Distortion

Welding distortion, a consequence of thermal contraction and solidification shrinkage during the laser beam welding (LBW) process, is a critical factor influencing the manufacturing of components [42]. The deformation caused by welding has a direct impact on the assembly and quality of products, with particular significance for thin-plate structures. Accurate prediction and understanding of welding distortion are essential for optimizing the design and manufacturing processes.

Several studies have delved into modeling welding distortion in the context of LBW, recognizing its importance in various applications. The complexities involved in predicting distortion arise from the intricate interplay of factors such as material properties, heat input, and weld configuration. Distortion modeling aims to provide insights into the deformation patterns that occur in the workpiece, enabling manufacturers to anticipate and mitigate potential issues.

Table 5.4 outlines some notable studies that have contributed to the understanding and prediction of welding distortion in LBW. These studies employ different approaches, including numerical simulations, experimental validations, and empirical methods. The diversity in methodologies highlights the multifaceted nature of the distortion phenomenon and the need for comprehensive modeling strategies.

According to the insights provided in Table 5.4, a comprehensive overview of distortion models is presented, categorized based on monitoring, approach, and methods, along with the associated error of the method. In a notable study [70], the deformation of thin-plate low-carbon steel (Q235), featuring a thickness of 2.3 mm, was numerically modeled using the thermos-elastic-plastic three-dimensional finite element method.

The study considered two distinct theories, namely large deformation (Case A) and small deformation (Case B), to capture the thermomechanical behavior of the welding process [70]. Experimental verification involved a specimen with specific processing parameters, including a laser power of 2,400 W, welding speed of 1.8 m/min, shielding gas flow rate of 10 L/mm, and a focus length of 200 mm [70].

The outcomes of this investigation yielded valuable insights. Firstly, it was observed that longitudinal bending exhibited minimal effects, while transverse bending (angular distortion) was notably pronounced. Secondly, the study highlighted that the predictions of Case A, utilizing the large deformation theory, aligned more closely with the measured values in terms of magnitude when compared to Case B, which employed the small deformation theory. Consequently, for enhanced prediction accuracy, the study recommended the utilization of the large deformation theory to simulate the thermomechanical behavior of thin-plate laser beam welding processes.

To achieve rapid deformation predictions in laser-welded thin sheets, a study [136] investigated both a local solid model (with a length of 80 mm) and a global model based on the inherent strain theory. This theory encompasses the summation of plastic strain, thermal strain, creep strain, and strain induced by phase transformation [137]. The inherent deformation schematic, including in-plane shrinkage and longitudinal bending, is detailed in the study [136].

The study employed an FEM-based model for thermos-elastic-plastic analysis, with the procedural details provided in the same study [136]. Experimental validation involved stainless steel 301, featuring a thickness of 1.33 mm, laser powers of 1,500 and 1,600 W, and welding speeds of 1.2 and 2 m/min. Altimeter measurements were used to quantify the out-of-plane welding deformation. Notably, the study reported a maximum deflection of 15 mm, exceeding ten times the plate thickness, at a welding speed of 2 m/min. Geometrical imperfections, with initial shapes of –10, 0, and +10, were applied to the plate, and the resulting curvature range was calculated [136].

The study's conclusion emphasized that the initial curvature of the plate, whether positive (convex shape) or negative (concave shape), directly influenced the final deformation shapes during the laser beam welding process. Convex and concave shapes were observed for positive and negative curvatures, respectively.

Examining Table 5.4 yields several key findings regarding distortion modeling. Firstly, the majority of distortion models are categorized as offline in terms

TABLE 5.4
Summary of Presented Studies Based on Weld Distortion, Types of Modeling, and Error

					Types of Modeling Based on				Max		
Row	Year	Material	Thickness (mm)	Max Temp (K)	Monitoring	Approach	Method	Types of Distortion	Distortion (mm)	Error (%)	References
1	2013	NR	3	NR	Off Post	Num Reg	ANSYS ANOVA	Out-of-plane deformation	0.52	NR	[138]
2	2014	Q235 Steel	2.3	2873.15	Off	Num	FEM	Transverse shrinkages, Longitudinal shrinkage, Out-of-plane deformation	8	NR	[70]
3	2016	SS 316L	5	1723	Off	Num	FEM	Angular deformation	2.5	5.95	[139]
4	2016	SS301	1.33	3487	Off	Num	FEM	Buckling distortion	15	5	[136]
5	2016	Q235 Steel	6	3573	Off	Num	FEM	Longitudinal deformation	0.17	NR	[72]
6	2017	316L	3.8	2783.15	Off	Num	FEM	Angular deformation	0.58	23.4	[140]
7	2017	SS301L	1.33	2123	Off	Num	ANSYS	Transverse/longitudinal deformation	4	NR	[141]
8	2019	SS316L	5	3403	Off	Num	ANSYS	Angular deformation	0.46	5.84	[142]
9	2021	DP 600 Steel	1.6	2523	Off	Num	SYSWELD	Transverse deformation	15	NR	[57]
10	2021	SS 304-AA6082 T6	1.5	723.15	Off	Num	ANSYS	Shrinkage/ Angular deformation	1.2	38.7	[143]
11	2021	SS 316LN	5.6	2068	Off	Num	SYSWELD	Longitudinal deformation	0.24	NR	[144]
12	2022	SS301L	0.6	2573	Off	Num	FEM	Transverse/longitudinal deformation	2.5	9.41	[124]
13	2022	SS316L	10	1673.15	Off	Num	SYSWELD	Angular distortion	0.47	14.35	[126]
14	2023	AA 5024	1.6	928	Off	Num	Simufact welding	Transverse/longitudinal deformation	1.05	NR	[145]

of monitoring, utilizing a numerical approach with commercial FEM-based software such as ANSYS and SYSWELD. These models address various types of distortions. Secondly, online analytical models have not considered previous distortion models, with only one post-process model based on regression (ANOVA) reporting no errors. Thirdly, the maximum error observed in distortion models ranges from 5% to 38.7%, remaining below 40%. This error pertains to BM thicknesses spanning from 0.6 to 10 mm and distortions ranging from 0.24 to 15 mm. These findings suggest the need for further studies to reduce the maximum error of models, thereby enhancing the accuracy of distortion predictions.

5.4 METALLURGICAL CHARACTERISTICS MODELING CASE STUDIES

In the upcoming subsections, we will explore laser beam welding (LBW) modeling focused on specific metallurgical characteristics, namely solidification mode, phase transformation, and morphology. Each of these characteristics will be introduced and discussed extensively, providing insights into recent developments in the field. These case studies will illuminate the progress made in LBW modeling of metallurgical characteristics, with a specific emphasis on solidification mode, phase transformation, and morphology. The incorporation of thermodynamic models, phase field simulations, and advanced computational techniques has played a pivotal role in advancing our understanding of microstructural evolution during LBW. These models empower researchers and engineers to optimize welding parameters, exert control over microstructural characteristics, and enhance the overall performance of LBW joints.

5.4.1 Solidification Mode and Phase Transformations

During the solidification process of a weld, a mushy zone is formed, encompassing both solid and liquid phases. This zone experiences tensile strain arising from shrinkage, thermal contraction, and resistance from the cooler BM [146]. The presence of this strain increases the risk of solidification cracking [147]. A visual representation of the mushy zone, including axial grains, partially melted metal, and columnar grains, is illustrated in detail in the study [150].

In the context of solidification cracking, the mushy zone plays a crucial role. The occurrence of cracking is mitigated when the flow of the interdendritic liquid compensates for local deformation. In simpler terms, the solidification crack is healed when there is sufficient liquid filling the space between deformed dendrites [148,149]. Therefore, it becomes evident that the solidification mode and phase transformations, influencing the characteristics of the mushy zone, are critical factors impacting solidification cracking, a significant criterion for evaluating weld performance [150].

In the context of pure metal solidification, the solid/liquid interface typically adopts a planar configuration unless the metal undergoes abrupt supercooling. In contrast, the solidification of alloys results in the breaking of the solid/liquid interface into cellular or dendritic structures. The specific structure formed is contingent upon the solidification conditions and the material composition. A comprehensive classification of solidification modes, as outlined in the study [151], includes four fundamental types: planar, cellular, columnar dendritic, and equiaxed dendritic modes [147].

While significant efforts have been directed towards modeling metallurgical characteristics in laser beam welding (LBW), real-time observation of the solidification process during the operation is challenging due to limitations in the monitoring process. Consequently, recent studies have predominantly relied on numerical methods for modeling. A selection of these numerical modeling approaches is summarized in Table 5.5. Primary dendritic arm spacing and secondary dendritic arm spacing have been reported to be related to the Gibbs-Thomson and curvature radius (R), based on the work of Kurz and Fisher [152]. The relationship is expressed as follows:

$$\lambda = A \times G_T^{-n} \times R^{-m} \tag{5.8}$$

In exploring the microstructural intricacies of the mixing of steel and aluminum during the LBW process, researchers have examined the coefficients A, n, and m, with various studies focusing on their implications. A noteworthy investigation combined numerical and EB methods to model the microstructure of steel and aluminum mixing [143]. The validity of the models was tested using stainless steel 304 and 6082-T6 aluminum, both with a thickness of 1.5 mm. Laser power was set at 3,750 Wa, and welding speeds of 4.2, 4.8, and 5.4 m/min were employed.

The weld regions were differentiated into zones A and B, corresponding to steel and aluminum parts, respectively, as detailed in the study [143]. The models aimed to predict the average aluminum concentrations within welds at different penetration depths (240, 320, 500, and 800 μm). The comparison between calculated and measured average aluminum concentrations, as well as the correlation of aluminum concentration across the entire weld (Al_W) to the average concentrations in zones A (Al_A) and B (Al_B), were presented in the study [143].

Empirical relations derived from the study allowed the estimation of the average aluminum concentration in zones A and B as $Al = 0.82 \times Al_W$ and $Al_B = (0.05 \times (A_{st}/A_{al}) + 1.15) \times Al_W$. This finding underscored a significant disparity in the aluminum concentration between the upper and lower zones of the St-Al welds. Considering the thermal expansion coefficient and elastic-plastic properties of the weld metal as functions of the aluminum concentration, the computational model should incorporate both zones for accurate predictions [143].

The exploration of weld microstructure modeling in LBW has yielded insights into various materials, primarily focusing on different steel types and Al alloys, characterized by thicknesses ranging from 1.5 to 10 mm. The models' foundation predominantly lies in offline monitoring processes, adopting numerical approaches with a reported error range of 5.5%–22.22%. It is noteworthy that a post-process-based regression model stands out with a distinct error characteristic.

TABLE 5.5
Summary of Presented Studies Based on Weld Microstructure and Types of Modeling and Error

			CR_{Max}	Thickness	Types of Modeling Based on				Max		
Row	Year	Material	(K/s)	(mm)	Monitoring	Approach	Method	Formula	Phases	Error (%)	References
1	2016	Q235 Steel	NR	6	Off	Num	FEM	NA	NR	15	[72]
2	2021	SS A387	4000	10	Off	Num	SYSWELD	Ms (°C)=539 – 423C – 30.4Mn – 17.7Ni – 12.1Cr – 7.5Mo + 10Co – 7.5Si	NR	5.5	[153]
3	2021	AH36 steel	NR	6	Off	Num	CFD-FEM	NA	NR	22.22	[154]
4	2021	DP 600 Steel	NR	1.6	Off	Num	SYSWELD	NA	Martensite 1%	NR	[57]
5	2021	SS 304-Al 6082-T6	1273	1.5	Post/Off	Reg/ Num	ANOVA/ ANSYS	$Al_B/Al_W=0.05\ (A_{st}/A_{al})+1.15$ $\alpha=18.3+0.083\ Al$	NR	NR	[143]

The numerical models invariably employ finite element methods and software platforms like SYSWELD, CFD, and ANSYS for accurate simulations.

Table 5.5 encapsulates the summary of these studies, offering a comprehensive overview of the types of modeling, weld microstructure considerations, and associated errors. The materials chosen for microstructure analysis, the numerical methodologies applied, and the software platforms utilized are systematically presented for a holistic understanding of the research landscape. Intriguingly, while the maximum reported error remains below 30%, a notable observation is the scarcity of closed-packed formulas in the majority of models, barring exceptions like studies [143,153]. This discrepancy highlights a prevailing challenge in LBW research—the absence of readily applicable closed-pack online formulas for predicting weld microstructure and phases.

5.4.2 Morphology

Weld morphology in the laser beam welding (LBW) process is intricately linked to the interaction between the laser beam and the molten pool, particularly influencing dendrite arm spacing (DAS) [150]. Various studies, employing diverse models, have ventured into the modeling and simulation of DAS in the LBW process. Table 5.6 encapsulates a subset of these studies, categorizing them based on monitoring and approach types, reported errors, and details of the LBW process.

The solidification of the melt pool in the LBW process is a nonlinear phenomenon, and the transient condition is considered to be closer to steady-state compared to the natural solidification process [155]. In a notable study [155], the phase field (PF) and cellular automata (CA) methods were employed to predict dendrite growth during the solidification process of LBW. The model's accuracy was assessed against experimental results, revealing a maximum error of 25% for DAS in the Al-Cu alloy across three different cases.

These investigations showcase the multi-faceted approaches undertaken to understand and predict weld morphology in LBW. The emphasis on DAS elucidates the importance of characterizing the microstructural features influenced by the complex interplay between laser interaction and solidification dynamics.

In the simulated microstructure of case A, as depicted in the study [155], the presence of a high density of initial seeds leads to robust diffusion field interactions among neighboring dendrites. This condition fosters competitive growth, resulting in certain grains surviving the competition while obstructing the growth of others. The utilization of the Phase Field (PF) model, as explored in additional studies [156–158], further contributes to the comprehensive understanding of microstructural evolution during the solidification process in laser beam welding.

This simulation not only provides insights into the intricate dynamics of competitive grain growth but also underlines the significance of initial seed density and diffusion field interactions in shaping the ultimate microstructure. The application of the PF model across various studies signifies its efficacy in capturing

TABLE 5.6
Summary of Presented Studies Based on Weld Morphology and Types of Modeling and Error

Row	Year	Material	CR_{Max} (K/s)	Thickness (mm)	Types of Modeling Based on			Formula	Max		References
					Monitoring	Approach	Method		GS (μm)	Error	
1	2011	Al-Cu	3,300	NR	Off	Num	PF-CA	NR	6.25	25	[155]
2	2017	Al Cu 2024	NR	2	Off	Num	PF	NR	10.8	97.8	[156]
3	2017	Ti-6Al-4V	410	1	Off	Num	ABAQUS	$d=54.82+1.86\times E_l^{a}$	1,000	1.1	[159]
4	2018	Al Cu 2024	NR	2	Off	Num	PF	NR	8.3	52.01	[157]
5	2021	Al4%Cu alloy 2A12	NR	4	Off	Num	PF	–	10	NR	[158]
6	2022	SS 316 NL	6,900	5.5	Off	Num	ANSYS	$SDAS=25\ CR^{-0.28}$	NR	NR	[97]

[a] E_l=Laser power/laser scanning speed.

the nuanced phenomena associated with dendritic growth and solidification in the context of LBW.

The microstructure images captured in proximity to the top surface of the melt pool during the laser welding process, conducted at a laser power of 2,500 W and welding velocity of 2.5 m/min, along with the corresponding Phase Field (PF) model representations, are prominently featured in the study [158]. Analyzing this from the mentioned study, one can deduce the model's success in accurately predicting the evolutionary dynamics of dendrite growth along the fusion boundary in the laser beam welding process. This observation underscores the reliability and effectiveness of the PF model in capturing the intricate details of microstructural changes during the solidification process in laser welding. The comparison between the experimental microstructure images and the PF model outputs serves as a testament to the model's capability in simulating and understanding dendritic growth phenomena under specific process conditions.

A comprehensive phase field (PF) model was developed to simulate the columnar-to-equiaxed transition (CET) throughout the entire melt pool of Al-4%wt-Cu alloy 2A12 during the laser beam welding (LBW) process, as detailed in the study [158]. The model incorporates considerations for crystalline orientations and heterogeneous nucleation, aligning with the specific experimental conditions involving a thickness of 4 mm, laser power of 3,000 W, welding speed of 3 mm/s, and a defocusing parameter of +10 mm. Electron backscatter diffraction was employed to observe the microstructure of the FZ.

Simulation outcomes highlighted the initiation of crystals as planar structures at the molten pool edge, transforming into columnar dendrites during the solidification process. Notably, dendrites exhibited growth toward the center of the FZ, regardless of their initial crystalline orientations. Equiaxed grains were observed to lead the growth, forming a belt ahead of columnar dendrites and impeding their progression. The top edge of the molten pool's cross-section demonstrated the highest concentration of equiaxed grains due to the fastest pulling velocity [158]. In-depth comparisons between the simulated and experimental results were conducted, encompassing various aspects such as the microstructure of the FZ, microstructure of the equiaxed grain zone, microstructure at the center of the equiaxed grain zone, grain size distribution of the equiaxed grain zone, and grain size distribution at the center of the equiaxed grain zone, offering valuable insights into the accuracy and reliability of the PF model.

Analyzing the models presented in Table 5.6, it is evident that all the showcased approaches are grounded in offline methodologies and utilize numerical techniques, encompassing phase field, cellular automata, and finite element methods. The predicted grain sizes exhibit a broad spectrum, spanning from 6.25 to 1,000 μm, with associated errors ranging from 1.1% to 97.8%. Diverse materials, including steel, Cu alloy, and Al alloy, characterized by thicknesses ranging from 1 to 5 mm, have been investigated within these models. An essential parameter for the modeling process, the maximum cooling rate, varies significantly and falls within the range of 410 to 6,900 K/s.

5.5 MECHANICAL CHARACTERISTICS MODELING CASE STUDIES

Mechanical characteristics, encompassing strength, residual stress, and hardness, stand as pivotal determinants influencing the performance and reliability of weld joints. The precise prediction and modeling of these mechanical characteristics during the laser beam welding (LBW) process hold critical importance for the design and assessment of welded structures. In this section, we will delve into recent studies dedicated to modeling the mechanical aspects of the LBW process, with a specific focus on strength, hardness, and residual stress.

These recent investigations showcase notable progress in the modeling of mechanical characteristics inherent in the LBW process. The incorporation of advanced techniques, such as finite element analysis, multi-physics models, and experimental validation, has significantly enriched our comprehension of the intricate relationships among process parameters, microstructure, and mechanical properties. Through the accurate prediction and modeling of these mechanical characteristics, researchers and engineers gain the capability to optimize weld design, assess joint performance, and ensure the structural integrity of welded constructions.

5.5.1 Strength

Strength stands as a fundamental mechanical characteristic of laser-welded joints, serving as a direct reflection of welding quality. Numerous studies have delved into in-depth research on yield strength and UTS within laser beam welding process models (LBWPM), some of which are detailed in Table 5.7. The presented strength models are categorized based on stages, including post-process and offline, and methods, encompassing regression, machine learning, and numerical approaches.

response surface method [160,161], support vector regression [115], Kirging [162], and XGBoost [36] are examples of regression-based methods utilized to model the strength of laser-welded joints, showcasing their efficacy in capturing the underlying relationships. Within the machine learning realm, neural networks (NN), genetic algorithm (GA) [163], and artificial neural network (ANN) [115,164] methods have been applied, demonstrating their capacity for predicting strength in the laser beam welding process.

Numerical approaches have been instrumental in modeling yield strength and UTS of laser-welded joints, with various finite element-based software such as ABAQUS [159], ANSYS [143], and SYSWELD [57,153] being recommended. While numerous studies propose numerical modeling for the yield stress of the LBW process, it is highlighted that yield strength is intricately related to grain size [165]. The Hall-Petch relation plays a crucial role in understanding this relationship, and its formulation is expressed as follows:

$$\sigma_y = \sigma_0 + k\lambda^{-0.5} \tag{5.9}$$

TABLE 5.7
Summary of Presented Studies Based on Weld Strength and Types of Modeling and Error

Row	Year	Material	Thickness (mm)	Types of Modeling Based on: Monitoring	Approach	Method	Formula	Max Strength (MPa): UTS	YS	Max Error (%)	References
1	2008	SS 304	3	Post	Reg	RSM	$TS = -4433.90 + 9102.06 \times P - 19.9 \times S - 511.93 \times F + 13.86 \times PS + 565.94 \times PF - 4.32 \times SF - 3954.102 \times P^2$	692	NR	6.98	[160]
2	2008	AA5182	1.4	Post	ML	NN-GA	NR	284.9	NR	0.9	[163]
3	2017	Ti-6Al-4V	1	Off	Num	ABAQUS	$UTS = 90.4 + 68.7 \times d^{-1/2}$[a]	915	–	60	[159]
							$YS = 797.1 + 56.7 d^{-1/2}$	–	839.5	54	
4	2017	NR	NR	Post	Reg	SVR	NR	60	NR	2	[115]
					ML	ANN				2	
5	2021	Al-Li	2.5–4.5	Post	Reg	XGboost	NR	293.5	NR	0.8	[36]
6	2021	DP 600 Steel	1.6	Off	Num	SYSWELD	NR	636	373	NR	[57]
7	2021	SS A387	10	Off	Num	SYSWELD	NR	556	NR	5.5	[153]
8	2021	301L-MT	8	Post	Reg	RSM	$TS = 743.07 - 25.99 \times A + 117.11 \times B - 67.7 \times C - 12.72 \times AB - 32.54 \times AC + 17.83 \times BC - 18.34 \times A^2 - 79.54 \times B^2 - 35.9 C^2$[b]	777.326	NR	11.04	[161]

(Continued)

TABLE 5.7 (*Continued*)
Summary of Presented Studies Based on Weld Strength and Types of Modeling and Error

				Types of Modeling Based on				Max			
								Strength (MPa)			
Row	Year	Material	Thickness (mm)	Monitoring	Approach	Method	Formula	UTS	YS	Error (%)	References
9	2021	SS 304-AA6082T6	1.5	Off	Num	ANSYS	Yield Strength = 3,781 × exp≈(−0.05654 × Al) − 3773 × exp (−0.1137 × Al) Yield strength = 2.2 × hardness − 232	1200	NR	38.7	[143]
10	2021	SS301L-MT	8	Post	Reg	RSM	Tensile strength = $743 - 25.99A + 117.11B - 67.7C - 12.72AB - 32.54AC + 17.84BC - 18.34A^2 - 79.54B^2 - 35.9C^2$	777.326	NR	11.04	[37]
11	2022	Al 2024T4-Al2024O	2	Off	Num	FEM	$\sigma_Y = -0.0543465$ HV $+ 216.37314$[c]	450	344	5	[35]
12	2022	SS 301	NR	Post	ML	ANN	NR	NA	3.1	5	[164]
13	2022	Al 6061	2	Post	Reg	Kirging, RSM, RBF	NR	230	NA	9.93	[162]
14	2022	iPP	2	Post/Off	Reg/Num	ANOVA/ COMSOL	$\sigma_{0.5\%CB} = 10.78 - 0.01P - 0.006919S + 0.000959P^2$	9	NA	10.52	[127]

[a] Where d is average width of prior-β grains.

[b] A is welding speed, B = laser power, C is wire feed rate, AC is the parameter interaction, B^2 and C^2 are the second order effect.

[c] c and d are fitting parameters and σ_Y is the yield stress.

Where σ_y is yield strength, σ_0 and k are LBW process coefficients. It is also investigated that yield stress can be approximated to a linear relationship with Vickers hardness (HV) as follows [35,166–168].

$$\sigma_Y = c.HV + d \tag{5.10}$$

where c and d are LBW process coefficients. Some of the results are provided in the study [143,162,164,142].

The comparison of the strength modeling of the laser beam welding (LBW) process presented in Table 5.7 reveals several noteworthy observations. Firstly, there is a lack of online-based models in previous studies, with the majority being offline and post-process models. This absence of real-time online models could impact the adaptability and responsiveness of LBW Process Models (LBWPM) to dynamic changes during the welding process. The error range reported in these studies spans from 0.9% to 60%, representing a broad spectrum of accuracy. Such variability may affect the reliability and precision of LBWPM, emphasizing the need for enhanced modeling strategies.

Additionally, the presented models exhibit limitations in terms of material and thickness considerations. The models are specific to particular materials such as steel, titanium alloy, Al alloy, and polymer, with thicknesses ranging from 1 to 10 mm. This specificity implies that the applicability of these models may be constrained by the defined material and thickness parameters. As a result, these models may not be universally applicable to a wide range of materials and thicknesses, warranting caution when extending their use to different scenarios.

Furthermore, comparing post-process regression-based approaches, it is evident that different formulas are presented for various experimental conditions, including thickness and material types (as observed in studies [35,37,127,143,159,160]). This underscores the limitation of post-process regression models, as they are tailored to specific processing parameters. Consequently, changes in process parameters necessitate the development of novel models to accurately reflect the altered conditions. In summary, the findings emphasize the need for more inclusive and adaptable LBWPM, considering real-time online models, broader material applicability, and improved methodologies for handling changes in processing parameters.

5.5.2 Hardness

Hardness, a critical mechanical characteristic, is defined as the material's toughness and is closely related to other mechanical properties, including tensile strength. The concept of brittleness, referring to the tendency of a material to break even under small, applied forces or at specific angles or planes, is inherently linked to hardness. Experimental analyses have provided evidence that brittleness tends to increase with higher hardness, establishing a strong correlation coefficient, particularly notable in laser beam welding (LBW) processes where intermetallic compounds (IMCs) are formed [169,170,171]. Consequently, the modeling of hardness in LBW processes directly impacts weld joint performance by influencing factors such as brittleness, which, in turn, contributes to fracture behavior and tensile strength.

Given the pivotal role of hardness in influencing weld joint performance, a comprehensive summary of hardness models, categorized based on monitoring, approach, and methods, is presented in Table 5.7. The hardness models discussed in this context primarily fall into the categories of post-process and offline, based on their stage of application. These models employ regression and numerical methods to predict the hardness of laser-welded joints. Regression-based models, such as ANOVA, are utilized alongside numerical methods employing software tools like SYSWELD [57], ANSYS [143], CFD-FEM [154], and FEM [35] to develop models for predicting the hardness of laser-welded joints.

In the study conducted by Mansur et al., a thorough comparison of actual microstructure and hardness values is juxtaposed with the temperature profile and estimation of martensite fraction obtained through numerical simulation, as illustrated in the study [57]. Through a meticulous validation process, the FEM-based model demonstrates excellent agreement with actual macrographic images and microhardness profiles. The model reveals a noteworthy observation: the hardness in the FZ exceeds that of the BM by approximately 60%. Notably, the supercritical HAZ stands out with the highest hardness, attributed to its finely refined microstructure. This investigation specifically delves into the characteristics of Dual-Phase 600 material with a thickness of 1.6 mm.

The Vickers hardness comparison between experimental and simulated results for laser-welded AH36 steel, featuring a 6 mm thickness, is depicted in the study [154]. Employing an offline Computational Fluid Dynamics-Finite Element Method (CFD-FEM) model, the study conducted a comprehensive analysis. The model incorporates a fourth-order polynomial function of martensite fraction (Fm) for hardness calculation, as outlined in Table 5.8. The fitting coefficient, denoted by α, illustrates a consistent trend between the model and experimental outcomes.

It is crucial to note that the observed reduction in hardness within the FZ region is attributed to the dendrite structure resulting from the melting of the workpiece. However, it's important to highlight that the study did not consider the hardness at the HAZ due to the omission of its hardening mechanism in the modeled scenario.

The hardness of the St-Al weld metal, portrayed as a nonlinear function of aluminum concentration, is meticulously modeled utilizing a numerical approach, as showcased in the study [143]. This representation vividly illustrates the impact of aluminum concentration on weld hardness. The study delves into the intricate microstructure resulting from the mixing of steel and aluminum within the weld pool during the keyhole laser beam welding (LBW) process.

In the context of the overlapping laser-welded austenitic stainless steel 304-6082-T6 aluminum alloy, featuring a thickness of 1.5 mm, the model provides valuable insights. Notably, it highlights the St-Al weld metal's progressive increase in hardness with a corresponding elevation in aluminum concentration, ranging from 0% to 9%.

TABLE 5.8
Summary of Presented Studies Based on Weld Hardness and Types of Modeling and Error

Row	Year	Material	Thickness (mm)	Types of Modeling based on			Formula	Max		References
				Monitoring	Approach	Method		Hardness (HV)	Error (%)	
1	2021	Al 2024T4-Al2024O	2	Off	Num	FEM	$HV = -0.0195.\sigma + 117.22$[a]	145	5	[35]
2	2021	DP 600 Steel	1.6	Off	Num	SYSWELD	NR	500	NR	[57]
3	2021	Ti-Grade 5	2	Post	Reg	ANOVA	NR	329	5.00	[122]
4	2021	SS 304-AA6082 T6	1.5	Off	Num	ANSYS	$Hardness = 582.3 \times \exp(-0.006 \times Al) - 574.4 \times \exp(-0.32 \times Al)$	600	38.70	[143]
5	2021	AH36 steel	6	Off	Num	CFD-FEM	$Hardness = \sum_{i=0}^{4} \left(\alpha_i F_m{}^i \right)$	440	22.22	[154]

[a] σ is the residual stress.

The analysis of hardness in the Laser Beam Welding (LBW) process extends to aluminum alloys 2024-T4 and Al 2024-O, each possessing a thickness of 2 mm. This scrutiny is conducted through an offline thermal elastic-plastic finite element model, as elucidated in the study [35]. The comprehensive examination of hardness involves a multifaceted investigation into the complexities of the material behavior under LBW conditions.

Diverse numerical models, as showcased in the study [35] and detailed in Table 5.8, contribute to the collective understanding of LBW hardness dynamics. A noteworthy revelation from these studies posits that Vickers hardness (HV) can be effectively approximated through a linear relationship with residual stress. The proposed relationship, documented in the literature [35,172–176], underscores the interplay between hardness and residual stress, providing valuable insights for the optimization and understanding of LBW processes.

$$\mathrm{HV} = k.\bar{\sigma} + b \tag{5.11}$$

The exploration of hardness models in the laser beam welding (LBW) process, as presented in Table 5.8, reveals a predominant reliance on offline and numerical methodologies. Notably, study [122] stands out as an exception in employing a distinct approach. These models, diverse in their focus on materials such as steel, aluminum alloy, and titanium alloy, with thickness variations spanning 1.5–6 mm, collectively contribute to the understanding of LBW hardness dynamics.

The discernment of material composition and residual stress impact on weld joint hardness has led to the formulation of closed-packed formulas in studies [35,143], and [154]. These formulas encapsulate the intricate interplay between material composition, residual stress, and resulting hardness. However, it is apparent that further investigations are warranted, particularly in the realm of polymer materials, utilizing online and exact methodologies.

As the field advances, future models could encompass a more comprehensive spectrum of parameters, considering interactions with material composition, phase transformations, solidification modes, and various mechanical, metallurgical, and geometrical characteristics. This holistic approach promises to refine the accuracy and applicability of LBW hardness models, contributing to a more nuanced understanding of the welding process.

5.5.3 Residual stress

Residual stresses in welding can be broadly categorized into microscopic and macroscopic types, each with distinct origins and characteristics [177]. Microscopic residual stresses are typically associated with metallurgical factors [178] and are influenced by the processing of raw materials [179]. On the other hand, macroscopic residual stresses, which include mechanical and thermal stresses, are caused by the mechanical behavior of various components [180].

The creation and distribution of microscopic residual stresses at the microstructural level are intricately linked to the materials used and their processing.

Macroscopic residual stresses, on the other hand, arise from the welding process itself and are influenced by factors such as temperature-induced deformation, degrees of temperature inhibition, mechanical constraints, and local/non-uniform heating conditions [181]. The interplay of these factors during welding gives rise to residual stresses that can impact the overall performance and integrity of the welded structure. Understanding and modeling these residual stresses are crucial for optimizing welding processes and ensuring the durability of welded joints.

In the early 1970s, Ueda and Hibbit et al. [182,183] introduced thermoplastic-plastic finite element analysis (FEA) as a method to simulate the welding temperature field, residual stress, and deformation. In contemporary research, numerous studies leverage finite element analyses to simulate welding processes, enabling the examination of residual stress distribution and deformation in welded joints [184]. Despite the various numerical models proposed for the residual stress in the LBW process, as illustrated in the study [74] and detailed in Table 5.9, certain relationships and formulations have been established to describe residual stresses. These relationships provide insights into the factors influencing residual stresses and contribute to a better understanding of their distribution in welded joints.

Recent studies, including those by Zhao et al. [74], have employed sophisticated thermal-elastic-plastic finite element theories and inherent strain analysis (ISA) using ANSYS software. The incorporation of thermal-plastic analyses allows researchers to model not only transverse and longitudinal residual stresses but also to analyze the effects of grain size and orientation changes in the re-melting zone, as demonstrated in Zhao et al.'s work [74]. It is worth noting that while solid-state phase transformation might have an insignificant influence on residual stress in low-carbon steel, the consideration of phase changes becomes crucial in more complex materials and situations. The complexity of LBW processes necessitates ongoing research, pushing the boundaries of modeling techniques and refining our understanding of the intricate relationships between mechanical behavior, temperature variations, and the resulting residual stresses. Future models could benefit from considering additional parameters, such as phase transformations, to achieve a more accurate representation of residual stress distribution in welded joints. These endeavors collectively pave the way for more reliable and comprehensive numerical simulations of LBW processes, bridging the gap between theoretical predictions and real-world outcomes [74,185].

$$\sigma = -\frac{E}{2(1+\nu)}\cot\theta_0 \frac{\pi}{180}\frac{\partial(2\theta)}{\partial \sin^2\psi} \tag{5.12}$$

where σ is the calculated residual stress. E, ν, and θ_0 are the elastic modulus, the Poisson's ratio, and the diffraction angle in the unstressed state, respectively, which are determined by the physical properties of the material. $\partial(2\theta)/\partial \sin^2\psi$ reflects the variation of lattice strain in different ψ angular directions. Although for low-carbon steel, solid-state phase transformation has an insignificant influence on residual stress welding [186], in the study [70], the phase changes were

TABLE 5.9
Summary of Presented Studies Based on Weld Residual Stress and Types of Modeling and Error

				Types of Modeling based on				Max		
Row	Year	Material	Thickness	Monitoring	Approach	Method	Formula	Stress (MPa)	Error (%)	References
1	2014	Q235	2.3	Off	Num	FEM	NA	300	NR	[70]
2	2016	Q235 Steel	6	Off	Num	FEM	NA	200	2.6	[72]
3	2016	SS 316L	5	Off	Num	FEM	NA	452.3	5.95	[139]
4	2017	SS 316L	5	Off	Num	ANSYS	NA	396	11.0	[187]
5	2017	316L	3.8	Off	Num	ANSYS	NA	720	23.4	[140]
6	2017	SS301L	1.33	Off	Num	ABAQUS	NA	550	NR	[141]
7	2018	Steel EH36	4	Off	Num	ANSYS	NA	650	15.6	[117]
8	2019	SS316L	5	Off	Num	ANSYS	NA	400	5.84	[142]
9	2019	SS316L	10	Off	Num	ANSYS	NA	190	NR	[74]
10	2021	HT780 steel	20	Off	Num	FEM	NA	500	NR	[77]
11	2021	SS 316LN	5.6	Off	Num	SYSWELD	NA	400	NR	[144]
12	2021	DP 600	1.6	Off	Num	SYSWELD	NA	600	NR	[57]
13	2021	AH36 steel	6	Off	Num	CFD-FEM	NA	600	75	[154]
14	2022	SS301L	0.6	Off	Num	FE	-	690.08	9.4	[124]
15	2022	Al 2024T4-Al2024O	2	Off	Num	FEM	Von Mises	300	5	[35]
16	2022	SS316L	10	Off	Num	SYSWELD	NA	452	5	[126]
17	2023	SS316L	5	Off	Num	ANSYS	NA	214	10	[188]
18	2023	Steel 710	6	Off/Post	Num/Reg	FEM/RFA	RS[a]	936.47	3.1	[135]

[a] $635.74+121.00A+62.96B-27.73C-24.69D+9.11AB-340.31AC-12.36AD+115.19BC-68.54BD+83.16CD+4.97A^2+28.11B^2+33.13C^2-37.42D^2$.

not taken into account in the model. It is also mentioned that the creep behavior was neglected because the period with high temperatures during the entire thermal cycle was very short. Hence, the total strain increments are decomposed into three components, including elastic strain (ε^e), plastic strain (ε^p), and thermal strain (εth), as follows:

$$\{d\varepsilon^{\text{total}}\} = \{d\varepsilon^e\} + \{d\varepsilon^p\} + \{d\varepsilon^{th}\} \tag{5.13}$$

$$\varepsilon_x = \frac{\partial u}{\partial x} + \frac{1}{2}\left\{\left(\frac{\partial u}{\partial x}\right)^2 + \left(\frac{\partial v}{\partial x}\right)^2 + \left(\frac{\partial w}{\partial x}\right)^2\right\} \tag{5.14}$$

$$\varepsilon_y = \frac{\partial v}{\partial y} + \frac{1}{2}\left\{\left(\frac{\partial u}{\partial y}\right)^2 + \left(\frac{\partial v}{\partial y}\right)^2 + \left(\frac{\partial w}{\partial y}\right)^2\right\} \tag{5.15}$$

$$\varepsilon_z = \frac{\partial w}{\partial z} + \frac{1}{2}\left\{\left(\frac{\partial u}{\partial z}\right)^2 + \left(\frac{\partial v}{\partial z}\right)^2 + \left(\frac{\partial w}{\partial z}\right)^2\right\} \tag{5.16}$$

$$\gamma_{xy} = \frac{\partial u}{\partial y} + \frac{\partial v}{\partial x} + \left\{\left(\frac{\partial u}{\partial x}\right)\left(\frac{\partial u}{\partial y}\right) + \left(\frac{\partial v}{\partial x}\right)\left(\frac{\partial v}{\partial y}\right) + \left(\frac{\partial w}{\partial x}\right)\left(\frac{\partial w}{\partial y}\right)\right\} \tag{5.17}$$

$$\gamma_{yz} = \frac{\partial v}{\partial z} + \frac{\partial w}{\partial y} + \left\{\left(\frac{\partial u}{\partial y}\right)\left(\frac{\partial u}{\partial z}\right) + \left(\frac{\partial v}{\partial y}\right)\left(\frac{\partial v}{\partial z}\right) + \left(\frac{\partial w}{\partial y}\right)\left(\frac{\partial w}{\partial z}\right)\right\} \tag{5.18}$$

$$\gamma_{zx} = \frac{\partial u}{\partial z} + \frac{\partial w}{\partial x} + \left\{\left(\frac{\partial u}{\partial z}\right)\left(\frac{\partial u}{\partial x}\right) + \left(\frac{\partial v}{\partial z}\right)\left(\frac{\partial v}{\partial x}\right) + \left(\frac{\partial w}{\partial z}\right)\left(\frac{\partial w}{\partial x}\right)\right\} \tag{5.19}$$

where ε_x, ε_y, and ε_z are Green-Lagrange strains in the x, y, and z directions. The shear strain on the x-y, y-z, and z-x planes is considered as γ_{xy}, γ_{yz}, and γ_{zx}, respectively. u, v, and w express the displacement in the x, y, and z directions, respectively. In equations (5.12)–(5.19), the first-order terms indicate linear behavior, and the second-order terms provide nonlinear responses, such as buckling [70]. The results revealed no difference between the transverse residual distribution on the top and bottom surfaces, and the maximum absolute value is approximately 50 MPa.

In a study focusing on the thermal-plastic analysis of 316L stainless steel multi-layer welds, researchers, as detailed in the study [74], employed experimental measurements, particularly X-ray diffraction, alongside numerical modeling to assess transverse and longitudinal residual stresses. Study [74] presents the measured and model results of transverse and longitudinal residual stress, transverse residual stress contour, and residual stress along thickness directions, respectively. Notably, there is a distinct change in residual stress at the interlayer position, indicating a direct influence of grain size and orientation changes in the re-melting zone.

It is intriguing to note that the models used in studies [74,140] were similar, both based on the transient thermal-elastic-plastic finite element theory and inherent strain analysis (ISA), utilizing ANSYS software. This consistency in the modeling approach across different heat sources highlights the versatility and applicability of the chosen methodology. As the LBW process involves complex interactions and varying conditions, the shared model in these studies contributes valuable insights into understanding the nuanced effects of heat sources on residual stress distribution in multi-layer stainless steel welds. Such integrated experimental and numerical investigations play a crucial role in advancing our comprehension of residual stresses in laser-welded joints, aiding in the development of more precise and reliable modeling techniques for practical applications [74,140].

In a study examining the residual stress in the LBW process, a thermo-elastoplastic model incorporating isotropic and kinematic hardening was employed, as detailed in the study [126]. These models exhibit similar behavior for initial traction; however, once the material undergoes plastic deformation, distinctions emerge. The isotropic hardening model expands the elasticity domain, resulting in increased yield strength. Conversely, the kinematic hardening model displaces the center of the elastic domain, representing the Bauschinger effect, making it more suitable for cyclic loading scenarios. The comparison between experimental and numerical models for transversal and longitudinal residual stresses, considering both hardening modes, is illustrated in the study [126] for laser scanning speeds of 600, 1,600, and 2,600 mm/min.

Observing the study [126], a pronounced influence of hardening patterns is evident in the HAZ, where the material undergoes plasticization and experiences cyclic loading. It is noteworthy that as the laser scanning speed increases, the maximum residual stresses also rise, with a notable 100 MPa increase observed in longitudinal residual stress between scanning speeds of 600 and 2,600 mm/min. This finding underscores the significance of considering hardening patterns in the modeling of residual stresses, especially in regions affected by plasticization and cyclic loading, offering valuable insights into the intricacies of material behavior during the LBW process [126].

Comparing the various models for residual weld stress presented in Table 5.9, it can be inferred that, with the exception of the study [135], which adopts a post-process regression-based approach, the majority of models are based on offline numerical methods utilizing the FEM. The reported error range for these models spans from 2.6% to 23.4%. While the models are not limited to specific materials, encompassing various types of steel and aluminum alloys with thicknesses ranging from 1.33 to 20 mm, it is notable that a closed-packed formula is not widely available in the literature. Moreover, an identified gap in the current literature pertains to the absence of a comprehensive closed-form solution for predicting residual stress, particularly within the domain of offline dynamic models [135].

This emphasizes the need for further research to address these gaps and advance the understanding and predictive capabilities of residual stress models in LBW processes. Future studies could explore the development of closed-form solutions and more accurate online models that account for the dynamic nature of the welding process. These endeavors would contribute to refining the accuracy and reliability of predicting residual stress in laser beam welding applications, ultimately enhancing the overall understanding of material behavior and weld quality [189,190].

REFERENCES

[1] K.-M. Hong and Y. C. Shin, "Prospects of laser welding technology in the automotive industry: A review," *Journal of Materials Processing Technology*, vol. 245, pp. 46–69, 2017.

[2] J. Blackburn, "Laser welding of metals for aerospace and other applications," In: *Welding and Joining of Aerospace Materials*. Elsevier, pp. 75–108, 2012.

[3] J. Wang, X. Fu, L. Zhang, Z. Zhang, J. Liu, and S. Chen, "A short review on laser welding/brazing of aluminum alloy to steel," *The International Journal of Advanced Manufacturing Technology*, vol. 112, pp. 2399–2411, 2021.

[4] A. Sadeghian and N. Iqbal, "A review on dissimilar laser welding of steel-copper, steel-aluminum, aluminum-copper, and steel-nickel for electric vehicle battery manufacturing," *Optics & Laser Technology*, vol. 146, p. 107595, 2022.

[5] L. Subashini, K. P. Prabhakar, S. Ghosh, and G. Padmanabham, "Comparison of laser-MIG hybrid and autogenous laser welding of M250 maraging steel thick sections-understanding the role of filler wire addition," *The International Journal of Advanced Manufacturing Technology*, vol. 107, pp. 1581–1594, 2020.

[6] T. Jokinen, M. Karhu, and V. Kujanpää, "Welding of thick austenitic stainless steel using Nd: Yttrium-aluminum-garnet laser with filler wire and hybrid process," *Journal of Laser Applications*, vol. 15, no. 4, pp. 220–224, 2003.

[7] A. Simar, Y. Bréchet, B. De Meester, A. Denquin, C. Gallais, and T. Pardoen, "Integrated modeling of friction stir welding of 6xxx series Al alloys: Process, microstructure and properties," *Progress in Materials Science*, vol. 57, no. 1, pp. 95–183, 2012.

[8] O. Myhr, Ø. Grong, H. Fjær, and C. Marioara, "Modeling of the microstructure and strength evolution in Al-Mg-Si alloys during multistage thermal processing," *Acta Materialia*, vol. 52, no. 17, pp. 4997–5008, 2004.

[9] D. Bardel et al., "Coupled precipitation and yield strength modeling for non-isothermal treatments of a 6061 aluminium alloy," *Acta Materialia*, vol. 62, pp. 129–140, 2014.

[10] X.-j. Cao, M. Jahazi, J. Immarigeon, and W. Wallace, "A review of laser welding techniques for magnesium alloys," *Journal of Materials Processing Technology*, vol. 171, no. 2, pp. 188–204, 2006.

[11] K. Hao, G. Li, M. Gao, and X. Zeng, "Weld formation mechanism of fiber laser oscillating welding of austenitic stainless steel," *Journal of Materials Processing Technology*, vol. 225, pp. 77–83, 2015.

[12] N. T. Ninh and M. A. Wahab, "The effect of residual stresses and weld geometry on the improvement of fatigue life," *Journal of Materials Processing Technology*, vol. 48, no. 1-4, pp. 581–588, 1995.

[13] X. Sun, E. V. Stephens, and M. A. Khaleel, "Effects of fusion zone size and failure mode on peak load and energy absorption of advanced high strength steel spot welds under lap shear loading conditions," *Engineering Failure Analysis*, vol. 15, no. 4, pp. 356–367, 2008.

[14] B. L. da Silva, J. Ferreira, and J. Araújo, "Influence of notch geometry on the estimation of the stress intensity factor threshold by considering the Theory of Critical Distances," *International Journal of Fatigue*, vol. 42, pp. 258–270, 2012.

[15] P. Konjatić, D. Kozak, and N. Gubeljak, "The influence of the weld width on fracture behaviour of the heterogeneous welded joint," *Key Engineering Materials*, vol. 488, pp. 367–370, 2012.

[16] J. Morris Jr, "The influence of grain size on the mechanical properties of steel," 2001.

[17] X. Zhang, G. Mi, and C. Wang, "Microstructure and performance of hybrid laser-arc welded high-strength low alloy steel and austenitic stainless steel dissimilar joint," *Optics & Laser Technology*, vol. 122, p. 105878, 2020.

[18] R. Farrar and P. Harrison, "Acicular ferrite in carbon-manganese weld metals: An overview," *Journal of Materials Science*, vol. 22, pp. 3812–3820, 1987.

[19] S. Ohkita and Y. Horii, "Recent development in controlling the microstructure and properties of low alloy steel weld metals," *ISIJ International*, vol. 35, no. 10, pp. 1170–1182, 1995.

[20] D. Jin, C. Hou, and L. Shen, "Effect of welding residual stress on the performance of CFST tubular joints," *Journal of Constructional Steel Research*, vol. 184, p. 106827, 2021.

[21] Y.-H. Zhang, S. Smith, L. Wei, C. Johnston, and A. Stacey, "Measurement and modeling of residual stresses in offshore circumferential welds," In: *International Conference on Offshore Mechanics and Arctic Engineering*. American Society of Mechanical Engineers, Nantes, France, June 9–14, p. V003T03A007, vol. 55355, 2013.

[22] S. I. Talabi, O. B. Owolabi, J. A. Adebisi, and T. Yahaya, "Effect of welding variables on mechanical properties of low carbon steel welded joint," *Advances in Production Engineering & Management*, vol. 9, no. 4, pp. 181–186, 2014.

[23] J. Dowden, "The theory of laser materials processing," *Heat and Mass Transfer in Modern Technology*, vol. 119, pp. 95-128, 2009.

[24] J. Yang *et al.*, "Laser techniques for dissimilar joining of aluminum alloys to steels: A critical review," *Journal of Materials Processing Technology*, vol. 301, p. 117443, 2022.

[25] L. F. Gonçalves, F. M. Duarte, C. I. Martins, and M. C. Paiva, "Laser welding of thermoplastics: An overview on lasers, materials, processes and quality," *Infrared Physics & Technology*, vol. 119, p. 103931, 2021.

[26] S. R. Parimanik, T. R. Mahapatra, and D. Mishra, "A systematic literature review on laser welding of NiTi SMA," *Lasers in Manufacturing and Materials Processing*, vol. 10, pp. 77–117, 2022.

[27] J. Junke, X. Jihao, J. Chenghu, L. SHENG, R. Haolei, and X. Hongbo, "Laser welding process and strength enhancement of carbon fiber reinforced thermoplastic composites and metals dissimilar joint: A review," *Chinese Journal of Aeronautics*, vol. 36, no. 12, 2023.

[28] M. Norouzian, M. A. Elahi, and P. Plapper, "A review: Suppression of the solidification cracks in the laser welding process by controlling the grain structure and chemical compositions," *Journal of Advanced Joining Processes*, vol. 7, p. 100139, 2023.

[29] S. Kang, K. Lee, M. Kang, Y. H. Jang, and C. Kim, "Weld-penetration-depth estimation using deep learning models and multisensor signals in Al/Cu laser overlap welding," *Optics & Laser Technology*, vol. 161, p. 109179, 2023.
[30] A. Franco, L. Romoli, and A. Musacchio, "Modeling for predicting seam geometry in laser beam welding of stainless steel," *International Journal of Thermal Sciences*, vol. 79, pp. 194–205, 2014.
[31] J. Kozak, "Prediction of weld deformations by numerical methods-review," *Polish Maritime Research*, vol. 29, no. 1, pp. 97–107, 2022.
[32] M. Hashemzadeh, Y. Garbatov, and C. G. Soares, "Hybrid-laser welding-induced distortions and residual stresses analysis of large-scale stiffener panel," *Ocean Engineering*, vol. 245, p. 110411, 2022.
[33] A. Mohan, D. Ceglarek, P. Franciosa, and M. Auinger, "Numerical study of beam oscillation and its effect on the solidification parameters and grain morphology in remote laser welding of high-strength aluminium alloys," *Science and Technology of Welding and Joining*, pp. 1–10, 2023.
[34] J. Hargreaves et al., "Microstructural modeling and characterisation of laser-keyhole welded Eurofer 97," *Materials & Design*, vol. 226, p. 111614, 2023.
[35] S. Gao, S. Geng, P. Jiang, C. Han, and L. Ren, "Numerical study on the effect of residual stress on mechanical properties of laser welds of aluminum alloy 2024," *Optics & Laser Technology*, vol. 146, p. 107580, 2022.
[36] Z. Zhang, Y. Huang, R. Qin, W. Ren, and G. Wen, "XGBoost-based on-line prediction of seam tensile strength for Al-Li alloy in laser welding: Experiment study and modeling," *Journal of Manufacturing Processes*, vol. 64, pp. 30–44, 2021.
[37] L. Chen, T. Yang, Y. Zhuang, and W. Chen, "The multi-objective optimization modeling for properties of 301 stainless steel welding joints in ultra-narrow gap laser welding," *Welding in the World*, vol. 65, pp. 1333–1345, 2021.
[38] J. Frostevarg, "Factors affecting weld root morphology in laser keyhole welding," *Optics and Lasers in Engineering*, vol. 101, pp. 89–98, 2018.
[39] S. F. Nabavi, M. H. Farshidianfar, A. Farshidianfar, and B. Beidokhti, "Physical-based methodology for prediction of weld bead characteristics in the laser edge welding process," *Optik*, vol. 241, p. 166917, 2021.
[40] L. Wei, D. Dean, and H. Murakawa, "Measurement of inherent deformations in typical weld joints using inverse analysis (Part2) Prediction of welding distortion of large structure," *Transactions of JWRI*, vol. 34, no. 1, pp. 113–123, 2005.
[41] V. Villaret *et al.*, "Characterization of gas metal arc welding welds obtained with new high Cr-Mo ferritic stainless steel filler wires," *Materials & Design*, vol. 51, pp. 474–483, 2013.
[42] S. Kou, "Welding metallurgy," *New Jersey, USA*, vol. 431, no. 446, pp. 223–225, 2003.
[43] T. Abioye, T. Olugbade, and T. Ogedengbe, "Welding of dissimilar metals using gas metal arc and laser welding techniques: A review," *Journal of Emerging Trends in Engineering and Applied Sciences*, vol. 8, no. 6, pp. 225–228, 2017.
[44] K. Jackson, *Solidification*. American Society for Metals, 1971.
[45] H. Inoue, T. Koseki, S. Okita, and M. Fuji, "Solidification and transformation behaviour of austenitic stainless steel weld metals solidified as primary austenite: Study of solidification and subsequent transformation of Cr-Ni stainless steel weld metals (1st Report)," *Welding International*, vol. 11, no. 11, pp. 876–887, 1997.
[46] M. H. Farshidianfar, "Real-time closed-loop control of microstructure and geometry in laser materials processing," PhD, University of Waterloo, 2017.
[47] R. Saluja and K. Moeed, "The emphasis of phase transformations and alloying constituents on hot cracking susceptibility of type 304L and 316L stainless steel welds," *International Journal of Engineering Science and Technology*, vol. 4, no. 5, pp. 2206–2216, 2012.

[48] N. Suutala, "Effect of solidification conditions on the solidification mode in austenitic stainless steels," *Metallurgical Transactions A*, vol. 14, no. 1, pp. 191–197, 1983.
[49] M. Pujar, R. Dayal, T. Gill, and S. Malhotra, "Evaluation of microstructure and electrochemical corrosion behavior of austenitic 316 stainless steel weld metals with varying chemical compositions," *Journal of Materials Engineering and Performance*, vol. 14, no. 3, pp. 327–342, 2005.
[50] D. Kotecki and T. Siewert, "WRC-1992 constitution diagram for stainless steel weld metals: A modification of the WRC-1988 diagram," *Welding Journal*, vol. 71, no. 5, pp. 171–178, 1992.
[51] Z. Zhang, B. Li, W. Zhang, R. Lu, S. Wada, and Y. Zhang, "Real-time penetration state monitoring using convolutional neural network for laser welding of tailor rolled blanks," *Journal of Manufacturing Systems*, vol. 54, pp. 348–360, 2020.
[52] S. S. R. Singh, R. V. Praneeth, V. S. Sankalp, S. S. Sashank, and R. Karthikeyan, "Welding, mechanical properties and microstructure of different grades of austenitic stainless steels: A review," *Materials Today: Proceedings*, vol. 62, pp. 3675–3680, 2022.
[53] K. Masubuchi, *Analysis of Welded Structures: Residual Stresses, Distortion, and their Consequences*. Elsevier, 2013.
[54] S. Sirohi, S. M. Pandey, V. Tiwari, D. Bhatt, D. Fydrych, and C. Pandey, "Impact of laser beam welding on mechanical behaviour of 2.25 Cr-1Mo (P22) steel," *International Journal of Pressure Vessels and Piping*, vol. 201, p. 104867, 2023.
[55] A. B. Naik and A. C. Reddy, "Optimization of tensile strength in TIG welding using the Taguchi method and analysis of variance (ANOVA)," *Thermal Science and Engineering Progress*, vol. 8, pp. 327–339, 2018.
[56] M. Sokolov, A. Salminen, M. Kuznetsov, and I. Tsibulskiy, "Laser welding and weld hardness analysis of thick section S355 structural steel," *Materials & Design*, vol. 32, no. 10, pp. 5127–5131, 2011.
[57] V. M. Mansur, R. A. de Figueiredo Mansur, S. M. de Carvalho, R. H. M. de Siqueira, and M. S. F. de Lima, "Effect of laser welding on microstructure and mechanical behaviour of dual phase 600 steel sheets," *Heliyon*, vol. 7, no. 12, p. e08601, 2021.
[58] Z. Barsoum and B. Jonsson, "Influence of weld quality on the fatigue strength in seam welds," *Engineering Failure Analysis*, vol. 18, no. 3, pp. 971–979, 2011.
[59] W. Cai, J. Wang, P. Jiang, L. Cao, G. Mi, and Q. Zhou, "Application of sensing techniques and artificial intelligence-based methods to laser welding real-time monitoring: A critical review of recent literature," *Journal of Manufacturing Systems*, vol. 57, pp. 1–18, 2020.
[60] S. Kaierle, M. Ungers, C. Franz, S. Mann, and P. Abels, "Understanding the laser process: New approaches for process monitoring in laser materials processing," vol. 7, pp. 49–52, 2010.
[61] Y. Kawahito, T. Ohnishi, and S. Katayama, "In-process monitoring and feedback control for stable production of full-penetration weld in continuous wave fibre laser welding," *Journal of Physics D: Applied Physics*, vol. 42, no. 8, p. 085501, 2009.
[62] D. You, X. Gao, and S. Katayama, "WPD-PCA-based laser welding process monitoring and defects diagnosis by using FNN and SVM," *IEEE Transactions on Industrial Electronics*, vol. 62, no. 1, pp. 628–636, 2014.
[63] A. Otto and M. Schmidt, "Towards a universal numerical simulation model for laser material processing," *Physics Procedia*, vol. 5, pp. 35–46, 2010.
[64] M. Geiger, K.-H. Leitz, H. Koch, and A. Otto, "A 3D transient model of keyhole and melt pool dynamics in laser beam welding applied to the joining of zinc coated sheets," *Production Engineering*, vol. 3, no. 2, pp. 127–136, 2009.

[65] K. Abderrazak, W. B. Salem, H. Mhiri, G. Lepalec, and M. Autric, "Modeling of CO_2 laser welding of magnesium alloys," *Optics & Laser Technology*, vol. 40, no. 4, pp. 581–588, 2008.
[66] M. M. Francois *et al.*, "Modeling of additive manufacturing processes for metals: Challenges and opportunities," *Current Opinion in Solid State and Materials Science*, vol. 21, no. 4, pp. 198–206, 2017.
[67] Sub-Projects Structure, Lehrstuhl für Photonische Technologien. https://www.for5134.science/en/teilprojekte/.
[68] A. Duggirala, P. Kalvettukaran, B. Acherjee, and S. Mitra, "Numerical simulation of the temperature field, weld profile, and weld pool dynamics in laser welding of aluminium alloy," *Optik*, vol. 247, p. 167990, 2021.
[69] A. H. Faraji, C. Maletta, G. Barbieri, F. Cognini, and L. Bruno, "Numerical modeling of fluid flow, heat, and mass transfer for similar and dissimilar laser welding of Ti-6Al-4V and Inconel 718," *The International Journal of Advanced Manufacturing Technology*, vol. 114, no. 3, pp. 899–914, 2021.
[70] J. Sun, X. Liu, Y. Tong, and D. Deng, "A comparative study on welding temperature fields, residual stress distributions and deformations induced by laser beam welding and CO2 gas arc welding," *Materials & Design*, vol. 63, pp. 519–530, 2014.
[71] C. Gobbi, "Low cost thermal imaging system for welding applications," Master of Applied Science, University of Waterloo, 2016.
[72] G. Mi, L. Xiong, C. Wang, X. Hu, and Y. Wei, "A thermal-metallurgical-mechanical model for laser welding Q235 steel," *Journal of Materials Processing Technology*, vol. 238, pp. 39–48, 2016.
[73] Z. Gao, P. Jiang, G. Mi, L. Cao, and W. Liu, "Investigation on the weld bead profile transformation with the keyhole and molten pool dynamic behavior simulation in high power laser welding," *International Journal of Heat and Mass Transfer*, vol. 116, pp. 1304–1313, 2018.
[74] L. Chen, G. Mi, X. Zhang, and C. Wang, "Numerical and experimental investigation on microstructure and residual stress of multi-pass hybrid laser-arc welded 316L steel," *Materials & Design*, vol. 168, p. 107653, 2019.
[75] E. Assuncao, S. Williams, and D. Yapp, "Interaction time and beam diameter effects on the conduction mode limit," *Optics and Lasers in Engineering*, vol. 50, no. 6, pp. 823–828, 2012.
[76] E. Assuncao and S. Williams, "Effect of material properties on the laser welding mode limits," *Journal of Laser Applications*, vol. 26, no. 1, p. 012008, 2014.
[77] F. Farrokhi, B. Endelt, and M. Kristiansen, "A numerical model for full and partial penetration hybrid laser welding of thick-section steels," *Optics & Laser Technology*, vol. 111, pp. 671–686, 2019.
[78] G. Wiklund, O. Akselsen, A. J. Sørgjerd, and A. F. Kaplan, "Geometrical aspects of hot cracks in laser-arc hybrid welding," *Journal of Laser Applications*, vol. 26, no. 1, p. 012003, 2014.
[79] M. Kristiansen, F. Farrokhi, E. Kristiansen, and S. Villumsen, "Application of hybrid laser arc welding for the joining of large offshore steel foundations," *Physics Procedia*, vol. 89, pp. 197–204, 2017.
[80] F. Farrokhi, S. E. Nielsen, R. Schmidt, S. Pedersen, and M. Kristiansen, "Effect of cut quality on hybrid laser arc welding of thick section steels," *Physics Procedia*, vol. 78, pp. 65–73, 2015.
[81] I. Bunaziv, J. Frostevarg, O. M. Akselsen, and A. F. Kaplan, "Process stability during fiber laser-arc hybrid welding of thick steel plates," *Optics and Lasers in Engineering*, vol. 102, pp. 34–44, 2018.

[82] U. Reisgen, S. Olschok, M. Weinbach, and O. Engels, "Welding of high thickness steel plates using a fiber coupled diode laser with 50 kW of output power," In: *Lasers in Manufacturing Conference*, München, Germany, 2017.

[83] Q. Pan, M. Mizutani, Y. Kawahito, and S. Katayama, "High power disk laser-metal active gas arc hybrid welding of thick high tensile strength steel plates," *Journal of Laser Applications*, vol. 28, no. 1, p. 012004, 2016.

[84] M. O. Gebhardt, A. Gumenyuk, V. Quiroz Penaranda, and M. Rethmeier, "Laser/GMA hybrid welding of thick-walled precision pipes," *Welding and Cutting*, vol. 11, no. 5, pp. 312–318, 2012.

[85] X. Cao, P. Wanjara, J. Huang, C. Munro, and A. Nolting, "Hybrid fiber laser-Arc welding of thick section high strength low alloy steel," *Materials & Design*, vol. 32, no. 6, pp. 3399–3413, 2011.

[86] F. Farrokhi, R. M. Larsen, and M. Kristiansen, "Single-pass hybrid laser welding of 25 mm thick steel," *Physics Procedia*, vol. 89, pp. 49–57, 2017.

[87] F. Vollertsen, S. GrÜnenwald, M. Rethmeier, A. Gumenyuk, U. Reisgen, and S. Olschok, "Welding thick steel plates with fibre lasers and GMAW," *Welding in the World*, vol. 54, no. 3, pp. R62–R70, 2010.

[88] M. Rethmeier, S. Gook, M. Lammers, and A. Gumenyuk, "Laser-hybrid welding of thick plates up to 32 mm using a 20 kW fibre laser," *Quarterly Journal of the Japan Welding Society*, vol. 27, no. 2, pp. 74s–79s, 2009.

[89] G. Li, C. Zhang, M. Gao, and X. Zeng, "Role of arc mode in laser-metal active gas arc hybrid welding of mild steel," *Materials & Design*, vol. 61, pp. 239–250, 2014.

[90] S. Morgan and S. Williams, "Hybrid laser conduction welding," In: *55th Annual Assembly of International Institute of Welding*, Copenhagen, Denmark, 2002.

[91] Y. Kawahito, M. Mizutani, and S. Katayama, "High quality welding of stainless steel with 10 kW high power fibre laser," *Science and Technology of Welding and Joining*, vol. 14, no. 4, pp. 288–294, 2009.

[92] S. Ramasamy and C. Albright, "CO_2 and Nd-YAG laser beam welding of 5754-O aluminium alloy for automotive applications," *Science and Technology of Welding and Joining*, vol. 6, no. 3, pp. 182–190, 2001.

[93] W. M. Steen and J. Mazumder, *Laser Material Processing*. Springer, 2010.

[94] X. Xie, J. Zhou, and J. Long, "Numerical study on molten pool dynamics and solute distribution in laser deep penetration welding of steel and aluminum," *Optics & Laser Technology*, vol. 140, p. 107085, 2021.

[95] C. Yuce, F. Karpat, and N. Yavuz, "Investigations on the microstructure and mechanical properties of laser welded dissimilar galvanized steel-aluminum joints," *The International Journal of Advanced Manufacturing Technology*, vol. 104, no. 5, pp. 2693–2704, 2019.

[96] Y. Wang, P. Jiang, J. Zhao, and S. Geng, "Effects of energy density attenuation on the stability of keyhole and molten pool during deep penetration laser welding process: A combined numerical and experimental study," *International Journal of Heat and Mass Transfer*, vol. 176, p. 121410, 2021.

[97] A. K. Unni and V. Muthukumaran, "Modeling of heat transfer, fluid flow, and weld pool dynamics during keyhole laser welding of 316 LN stainless steel using hybrid conical-cylindrical heat source," *The International Journal of Advanced Manufacturing Technology*, vol. 122, no. 9, pp. 3623–3645, 2022.

[98] L. Zhang, J. Zhang, G. Zhang, W. Bo, and S. Gong, "An investigation on the effects of side assisting gas flow and metallic vapour jet on the stability of keyhole and molten pool during laser full-penetration welding," *Journal of Physics D: Applied Physics*, vol. 44, no. 13, p. 135201, 2011.

[99] F. Tenner, C. Brock, F.-J. Gürtler, F. Klämpfl, and M. Schmidt, "Experimental and numerical analysis of gas dynamics in the keyhole during laser metal welding," *Physics Procedia*, vol. 56, pp. 1268–1276, 2014.
[100] K. N. Lankalapalli, J. F. Tu, and M. Gartner, "A model for estimating penetration depth of laser welding processes," *Journal of Physics D: Applied Physics*, vol. 29, no. 7, p. 1831, 1996.
[101] C. Lampa, A. F. Kaplan, J. Powell, and C. Magnusson, "An analytical thermodynamic model of laser welding," *Journal of Physics D: Applied Physics*, vol. 30, no. 9, p. 1293, 1997.
[102] K. Lankalapalli, J. Tu, K. Leong, and M. Gartner, "Laser weld penetration estimation using temperature measurements," *Journal of Manufacturing Science and Engineering*, vol. 121, pp. 179–188, 1999.
[103] F.-R. Tsai and E. Kannatey-Asibu Jr, "Modeling of conduction mode laser welding process for feedback control," *Journal of Manufacturing Science and Engineering*, vol. 122, no. 3, pp. 420–428, 2000.
[104] H. Zhao and T. DebRoy, "Weld metal composition change during conduction mode laser welding of aluminum alloy 5182," *Metallurgical and Materials Transactions B*, vol. 32, no. 1, pp. 163–172, 2001.
[105] Y. W. Park, H. Park, S. Rhee, and M. Kang, "Real time estimation of CO2 laser weld quality for automotive industry," *Optics & Laser Technology*, vol. 34, no. 2, pp. 135–142, 2002.
[106] G. Casalino and F. M. C. Minutolo, "A model for evaluation of laser welding efficiency and quality using an artificial neural network and fuzzy logic," *Proceedings of the Institution of Mechanical Engineers, Part B: Journal of Engineering Manufacture*, vol. 218, no. 6, pp. 641–646, 2004.
[107] H. Du, L. Hu, J. Liu, and X. Hu, "A study on the metal flow in full penetration laser beam welding for titanium alloy," *Computational Materials Science*, vol. 29, no. 4, pp. 419–427, 2004.
[108] K. Benyounis, A. Olabi, and M. Hashmi, "Effect of laser welding parameters on the heat input and weld-bead profile," *Journal of Materials Processing Technology*, vol. 164, pp. 978–985, 2005.
[109] K. Balasubramanian, G. Buvanashekaran, and K. Sankaranarayanasamy, "Modeling of laser beam welding of stainless steel sheet butt joint using neural networks," *CIRP Journal of Manufacturing Science and Technology*, vol. 3, no. 1, pp. 80–84, 2010.
[110] D. Hann, J. Iammi, and J. Folkes, "Keyholing or conduction-prediction of laser penetration depth," In: *Proceedings of the 36th International MATADOR Conference*. Springer, pp. 275–278, 2010.
[111] M. Tobar, M. Lamas, A. Yáñez, J. Sánchez-Amaya, Z. Boukha, and F. Botana, "Experimental and simulation studies on laser conduction welding of AA5083 aluminium alloys," *Physics Procedia*, vol. 5, pp. 299–308, 2010.
[112] D. Hann, J. Iammi, and J. Folkes, "A simple methodology for predicting laser-weld properties from material and laser parameters," *Journal of Physics D: Applied Physics*, vol. 44, no. 44, p. 445401, 2011.
[113] M. Luo and Y. C. Shin, "Estimation of keyhole geometry and prediction of welding defects during laser welding based on a vision system and a radial basis function neural network," *The International Journal of Advanced Manufacturing Technology*, vol. 81, no. 1, pp. 263–276, 2015.
[114] Y. Zhang, X. Gao, and S. Katayama, "Weld appearance prediction with BP neural network improved by genetic algorithm during disk laser welding," *Journal of Manufacturing Systems*, vol. 34, pp. 53–59, 2015.

[115] D. Petković, "Prediction of laser welding quality by computational intelligence approaches," *Optik*, vol. 140, pp. 597–600, 2017.

[116] A. Artinov, M. Bachmann, and M. Rethmeier, "Equivalent heat source approach in a 3D transient heat transfer simulation of full-penetration high power laser beam welding of thick metal plates," *International Journal of Heat and Mass Transfer*, vol. 122, pp. 1003–1013, 2018.

[117] Y. Rong, G. Mi, J. Xu, Y. Huang, and C. Wang, "Laser penetration welding of ship steel EH36: A new heat source and application to predict residual stress considering martensite phase transformation," *Marine Structures*, vol. 61, pp. 256–267, 2018.

[118] Z. Lei, J. Shen, Q. Wang, and Y. Chen, "Real-time weld geometry prediction based on multi-information using neural network optimized by PCA and GA during thin-plate laser welding," *Journal of Manufacturing Processes*, vol. 43, pp. 207–217, 2019.

[119] B. Liu, W. Jin, A. Lu, K. Liu, C. Wang, and G. Mi, "Optimal design for dual laser beam butt welding process parameter using artificial neural networks and genetic algorithm for SUS316L austenitic stainless steel," *Optics & Laser Technology*, vol. 125, p. 106027, 2020.

[120] Z. Li, K. Rostam, A. Panjehpour, M. Akbari, A. Karimipour, and S. Rostami, "Experimental and numerical study of temperature field and molten pool dimensions in dissimilar thickness laser welding of Ti6Al4V alloy," *Journal of Manufacturing Processes*, vol. 49, pp. 438–446, 2020.

[121] M. S. Khan, S. Shahabad, M. Yavuz, W. Duley, E. Biro, and Y. Zhou, "Numerical modeling and experimental validation of the effect of laser beam defocusing on process optimization during fiber laser welding of automotive press-hardened steels," *Journal of Manufacturing Processes*, vol. 67, pp. 535–544, 2021.

[122] T. Le-Quang, N. Faivre, F. Vakili-Farahani, and K. Wasmer, "Energy-efficient laser welding with beam oscillating technique-A parametric study," *Journal of Cleaner Production*, vol. 313, p. 127796, 2021.

[123] S. F. Nabavi, M. H. Farshidianfar, A. Farshidianfar, and B. Beidokhti, "Physical-based methodology for prediction of weld bead characteristics in the Laser Edge Welding process," *Optik*, vol. 241, p. 166917, 2021.

[124] Z. Liu, X. Jin, J. Li, Z. Hao, and J. Zhang, "Numerical simulation and experimental analysis on the deformation and residual stress in trailing ultrasonic vibration assisted laser welding," *Advances in Engineering Software*, vol. 172, p. 103200, 2022.

[125] V. Busto, D. Coviello, A. Lombardi, M. De Vito, and D. Sorgente, "Thermal finite element modeling of the laser beam welding of tailor welded blanks through an equivalent volumetric heat source," *The International Journal of Advanced Manufacturing Technology*, vol. 119, no. 1, pp. 137–148, 2022.

[126] Y. Jia, Y. Saadlaoui, H. Hamdi, J. Sijobert, J.-C. Roux, and J.-M. Bergheau, "An experimental and numerical case study of thermal and mechanical consequences induced by laser welding process," *Case Studies in Thermal Engineering*, vol. 35, p. 102078, 2022.

[127] F. Dave, M. M. Ali, M. Mokhtari, R. Sherlock, A. McIlhagger, and D. Tormey, "Effect of laser processing parameters and carbon black on morphological and mechanical properties of welded polypropylene," *Optics & Laser Technology*, vol. 153, p. 108216, 2022.

[128] X. Fan, G. Qin, Z. Jiang, and H. Wang, "Comparative analysis between the laser beam welding and low current pulsed GMA assisted high-power laser welding by numerical simulation," *Journal of Materials Research and Technology*, vol. 22, pp. 2549–2565, 2023.

[129] W. Ke et al., "Heat transfer and melt flow of keyhole, transition and conduction modes in laser beam oscillating welding," *International Journal of Heat and Mass Transfer*, vol. 203, p. 123821, 2023.
[130] C. Tan *et al.*, "Numerical and experimental study of thermal fluid flow and keyhole dynamic in laser welding of aluminum alloy assisted by electromagnetic field," *Optics & Laser Technology*, vol. 157, p. 108718, 2023.
[131] J. Zhao, J. Wang, X. Kang, X. Wang, and X. Zhan, "Effect of beam oscillation and oscillating frequency induced heat accumulation on microstructure and mechanical property in laser welding of Invar alloy," *Optics & Laser Technology*, vol. 158, p. 108831, 2023.
[132] Y. Lu, Y. Deng, L. Shi, L. Jiang, and M. Gao, "Numerical simulation of thermal flow dynamics in oscillating laser welding of aluminum alloy," *Optics & Laser Technology*, vol. 159, p. 109003, 2023.
[133] N. T. Tien, Y.-L. Lo, M. M. Raza, C.-Y. Chen, and C.-P. Chiu, "Optimization of processing parameters for pulsed laser welding of dissimilar metal interconnects," *Optics & Laser Technology*, vol. 159, p. 109022, 2023.
[134] X. Zhao, J. Chen, W. Zhang, and H. Chen, "A study on weld morphology and periodic characteristics evolution of circular oscillating laser beam welding of SUS301L-HT stainless steel," *Optics & Laser Technology*, vol. 159, p. 109030, 2023.
[135] S. S. Surwase and S. P. Bhosle, "Investigating the effect of residual stresses and distortion of laser welded joints for automobile chassis and optimizing weld parameters using random forest based grey wolf optimizer," *Welding International*, vol. 37, pp. 46–67, 2023.
[136] H. Huang, J. Wang, L. Li, and N. Ma, "Prediction of laser welding induced deformation in thin sheets by efficient numerical modeling," *Journal of Materials Processing Technology*, vol. 227, pp. 117–128, 2016.
[137] H. Murakawa, D. Deng, and N. Ma, "Concept of inherent strain, inherent stress, inherent deformation and inherent force for prediction of welding distortion and residual stress," *Transactions of JWRI*, vol. 39, no. 2, pp. 103–105, 2010.
[138] J. W. Kim, B. S. Jang, Y. T. Kim, and K. San Chun, "A study on an efficient prediction of welding deformation for T-joint laser welding of sandwich panel PART I: Proposal of a heat source model," *International Journal of Naval Architecture and Ocean Engineering*, vol. 5, no. 3, pp. 348–363, 2013.
[139] Y. Rong, Y. Huang, G. Zhang, G. Mi, and W. Shao, "Laser beam welding of 316L T-joint: Microstructure, microhardness, distortion, and residual stress," *The International Journal of Advanced Manufacturing Technology*, vol. 90, no. 5, pp. 2263–2270, 2017.
[140] Y. Rong, Y. Huang, J. Xu, H. Zheng, and G. Zhang, "Numerical simulation and experiment analysis of angular distortion and residual stress in hybrid laser-magnetic welding," *Journal of Materials Processing Technology*, vol. 245, pp. 270–277, 2017.
[141] H. Huang, S. Tsutsumi, J. Wang, L. Li, and H. Murakawa, "High performance computation of residual stress and distortion in laser welded 301L stainless sheets," *Finite elements in analysis and design*, vol. 135, pp. 1–10, 2017.
[142] J. Xu, C. Chen, T. Lei, W. Wang, and Y. Rong, "Inhomogeneous thermal-mechanical analysis of 316L butt joint in laser welding," *Optics & Laser Technology*, vol. 115, pp. 71–80, 2019.
[143] A. Evdokimov, N. Doynov, R. Ossenbrink, A. Obrosov, S. Weiß, and V. Michailov, "Thermomechanical laser welding simulation of dissimilar steel-aluminum overlap joints," *International Journal of Mechanical Sciences*, vol. 190, p. 106019, 2021.

[144] M. Ragavendran and M. Vasudevan, "Effect of laser and hybrid laser welding processes on the residual stresses and distortion in AISI type 316L (N) stainless steel weld joints," *Metallurgical and Materials Transactions B*, vol. 52, no. 4, pp. 2582–2603, 2021.

[145] K. Barat, C. N. Kumar, G. Patil, K. Panbarasu, and K. Venkateswarlu, "Defectometry, distortion analysis and metallurgical properties in the keyhole and conduction mode of laser beam welding of Al-Mg-Sc alloy butt joints," *Optics & Laser Technology*, vol. 162, p. 109248, 2023.

[146] F. Bodaghi, M. Movahedi, and A. Kokabi, "Estimation of solidification cracking susceptibility in Al-Si-Cu alloy weld: Effects of anisotropic permeability and deformation orientation," *Journal of Materials Research and Technology*, vol. 23, pp. 2351–2361, 2023.

[147] S. Kou, "Solidification and liquation cracking issues in welding," *Jom*, vol. 55, pp. 37–42, 2003.

[148] J. Liu, Z. Rao, S. Liao, and P.-C. Wang, "Modeling of transport phenomena and solidification cracking in laser spot bead-on-plate welding of AA6063-T6 alloy. Part II-simulation results and experimental validation," *The International Journal of Advanced Manufacturing Technology*, vol. 74, pp. 285–296, 2014.

[149] T. Soysal and S. Kou, "Role of liquid backfilling in reducing solidification cracking in aluminium welds," *Science and Technology of Welding and Joining*, vol. 25, no. 5, pp. 415–421, 2020.

[150] L. Wang, M. Gao, C. Zhang, and X. Zeng, "Effect of beam oscillating pattern on weld characterization of laser welding of AA6061-T6 aluminum alloy," *Materials & Design*, vol. 108, pp. 707–717, 2016.

[151] K. Balasubramanian, S. Raghavendran, and V. Balusamy, "Studies on the effect of mechanical vibration on the microstructure of the weld metal," *International journal of technology and engineering systems*, vol. 2, no. 3, pp. 253–256, 2011.

[152] W. Kurz and D. Fisher, "Dendrite growth at the limit of stability: Tip radius and spacing," *Acta Metallurgica*, vol. 29, no. 1, pp. 11–20, 1981.

[153] Z. Du, X. Sun, F. L. Ng, Y. Chew, C. Tan, and G. Bi, "Thermo-metallurgical simulation and performance evaluation of hybrid laser arc welding of chromium-molybdenum steel," *Materials & Design*, vol. 210, p. 110029, 2021.

[154] S.-W. Han, W.-I. Cho, L.-J. Zhang, and S.-J. Na, "Coupled simulation of thermal-metallurgical-mechanical behavior in laser keyhole welding of AH36 steel," *Materials & Design*, vol. 212, p. 110275, 2021.

[155] W. Tan, N. S. Bailey, and Y. C. Shin, "A novel integrated model combining cellular automata and phase field methods for microstructure evolution during solidification of multi-component and multi-phase alloys," *Computational Materials Science*, vol. 50, no. 9, pp. 2573–2585, 2011.

[156] L. Wang *et al.*, "Simulation of dendrite growth in the laser welding pool of aluminum alloy 2024 under transient conditions," *Journal of Materials Processing Technology*, vol. 246, pp. 22–29, 2017.

[157] S. Geng *et al.*, "Cellular automaton modeling for dendritic growth during laser beam welding solidification process," *Journal of Laser Applications*, vol. 30, no. 3, p. 032406, 2018.

[158] X. Lingda, Z. Guoli, M. Gaoyang, W. Chunming, and J. Ping, "A phase-field simulation of columnar-to-equiaxed transition in the entire laser welding molten pool," *Journal of Alloys and Compounds*, vol. 858, p. 157669, 2021.

[159] Y. Ren, X. Lin, X. Fu, H. Tan, J. Chen, and W. Huang, "Microstructure and deformation behavior of Ti-6Al-4V alloy by high-power laser solid forming," *Acta Materialia*, vol. 132, pp. 82–95, 2017.

[160] K. Benyounis, A.-G. Olabi, and M. Hashmi, "Multi-response optimization of CO2 laser-welding process of austenitic stainless steel," *Optics & Laser Technology*, vol. 40, no. 1, pp. 76–87, 2008.
[161] L. Chen, T. Yang, Y. Zhuang, and W. Chen, "The multi-objective optimization modeling for properties of 301 stainless steel welding joints in ultra-narrow gap laser welding," *Welding in the World*, vol. 65, no. 7, pp. 1333–1345, 2021.
[162] Y. Li, M. Xiong, Y. He, J. Xiong, X. Tian, and P. Mativenga, "Multi-objective optimization of laser welding process parameters: The trade-offs between energy consumption and welding quality," *Optics & Laser Technology*, vol. 149, p. 107861, 2022.
[163] Y. W. Park and S. Rhee, "Process modeling and parameter optimization using neural network and genetic algorithms for aluminum laser welding automation," *The International Journal of Advanced Manufacturing Technology*, vol. 37, no. 9, pp. 1014–1021, 2008.
[164] N. Banerjee, A. Biswas, M. Kumar, A. Sen, and S. Maity, "Modeling of laser welding of stainless steel using artificial neural networks," *Materials Today: Proceedings*, vol. 66, pp. 1784–1788 2022.
[165] E. Hall, "The deformation and ageing of mild steel: III discussion of results," *Proceedings of the Physical Society. Section B*, vol. 64, no. 9, p. 747, 1951.
[166] R. Rodrıguez and I. Gutierrez, "Correlation between nanoindentation and tensile properties: Influence of the indentation size effect," *Materials Science and Engineering*: *A*, vol. 361, no. 1-2, pp. 377–384, 2003.
[167] M. Tiryakioğlu, "On the relationship between Vickers hardness and yield stress in Al-Zn-Mg-Cu Alloys," *Materials Science and Engineering: A*, vol. 633, pp. 17–19, 2015.
[168] M. Tiryakioğlu, J. Robinson, M. Salazar-Guapuriche, Y. Zhao, and P. Eason, "Hardness-strength relationships in the aluminum alloy 7010," *Materials Science and Engineering*: *A*, vol. 631, pp. 196–200, 2015.
[169] J. Zou, J. Han, and W. Yang, "Investigating the influences of indentation hardness and brittleness of rock-like material on its mechanical crushing behaviors," *Mathematical Problems in Engineering*, vol. 2020, pp. 1–16, 2020.
[170] J. Zhou, S. Mandal, F. Chen, M. Quest, and D. Hume, "Reservoir geomechanic heterogeneity index (RGHI): Concept, methodology, and application," In: *SPE/AAPG/SEG Unconventional Resources Technology Conference, 2018: OnePetro, Houston, Texas, USA*.
[171] A. S. Zoeram and S. A. Mousavi, "Laser welding of Ti-6Al-4V to Nitinol," *Materials & Design*, vol. 61, pp. 185–190, 2014.
[172] G. Sines and R. Carlson, "Hardness measurements for determination of residual stresses," *ASTM Bull*, vol. 180, pp. 35–37, 1952.
[173] T. Tsui, W. Oliver, and G. Pharr, "Influences of stress on the measurement of mechanical properties using nanoindentation: Part I. Experimental studies in an aluminum alloy," *Journal of Materials Research*, vol. 11, no. 3, pp. 752–759, 1996.
[174] A. Bolshakov, W. Oliver, and G. Pharr, "Influences of stress on the measurement of mechanical properties using nanoindentation: Part II. Finite element simulations," *Journal of Materials Research*, vol. 11, no. 3, pp. 760–768, 1996.
[175] M. Khan, M. Fitzpatrick, S. Hainsworth, and L. Edwards, "Effect of residual stress on the nanoindentation response of aerospace aluminium alloys," *Computational Materials Science*, vol. 50, no. 10, pp. 2967–2976, 2011.
[176] N. Huber and J. Heerens, "On the effect of a general residual stress state on indentation and hardness testing," *Acta Materialia*, vol. 56, no. 20, pp. 6205–6213, 2008.

[177] M. Sabokrouh and M. Farahani, "Correlation between the weld residual stresses and its tensile and impact strength," *Journal of Applied and Computational Mechanics*, vol. 5, no. 4, pp. 727–734, 2019.
[178] M. Farahani, I. Sattari-Far, D. Akbari, and R. Alderliesten, "Numerical and experimental investigations of effects of residual stresses on crack behavior in Aluminum 6082-T6," *Proceedings of the Institution of Mechanical Engineers, Part C: Journal of Mechanical Engineering Science*, vol. 226, no. 9, pp. 2178–2191, 2012.
[179] M. Farahani and I. Sattari-Far, "Effects of residual stresses on crack-tip constraints," *Scientia Iranica*, vol. 18, no. 6, pp. 1267–1276, 2011.
[180] J. Francis, H. Bhadeshia, and P. Withers, "Welding residual stresses in ferritic power plant steels," *Materials Science and Technology*, vol. 23, no. 9, pp. 1009–1020, 2007.
[181] S. H. Zargar, M. Farahani, and M. K. B. Givi, "Numerical and experimental investigation on the effects of submerged arc welding sequence on the residual distortion of the fillet welded plates," *Proceedings of the Institution of Mechanical Engineers, Part B: Journal of Engineering Manufacture*, vol. 230, no. 4, pp. 654–661, 2016.
[182] Y. Ueda and T. Yamakawa, "Analysis of thermal elastic-plastic stress and strain during welding by finite element method," *Japan Welding Society Transactions*, vol. 2, no. 2, 1971.
[183] H. D. Hibbitt and P. V. Marcal, "A numerical, thermo-mechanical model for the welding and subsequent loading of a fabricated structure," *Computers & Structures*, vol. 3, no. 5, pp. 1145–1174, 1973.
[184] Y. Zhang, W. Su, H. Dong, T. Li, and H. Cao, "Effect of Welding Sequence and Constraint on the Residual Stress and Deformation of Thick Welded Butt Joint Made of Q345qD Steel," *Advances in Civil Engineering*, vol. 2022, 2022.
[185] P. S. Prevey, "X-ray diffraction residual stress techniques," *ASM International, ASM Handbook.*, vol. 10, pp. 380–392, 1986.
[186] D. Deng, "FEM prediction of welding residual stress and distortion in carbon steel considering phase transformation effects," *Materials & Design*, vol. 30, no. 2, pp. 359–366, 2009.
[187] Y. Rong, G. Zhang, and Y. Huang, "Study on deformation and residual stress of laser welding 316L T-joint using 3D/shell finite element analysis and experiment verification," *The International Journal of Advanced Manufacturing Technology*, vol. 89, pp. 2077–2085, 2017.
[188] H. Vemanaboina et al., "Evaluation of residual stresses in CO2 laser beam welding of SS316L weldments using FEA," *Materials Research Express*, vol. 10, p. 016509 2023.
[189] F. Moglia and A. Raspa, "New trends in laser beam welding: How automotive applications are driving the future of laser technologies," *PhotonicsViews*, vol. 17, no. 50, pp. 26–29, 2020.
[190] T. Mo, Y. Li, K.-t. Lau, C. K. Poon, Y. Wu, and Y. Luo, "Trends and emerging technologies for the development of electric vehicles," *Energies*, vol. 15, no. 17, p. 6271, 2022.

6 Laser Beam Cladding

Types and Processing Parameters

6.1 INTRODUCTION

The combined effects of wear, corrosion, and fatigue pose a significant threat to the longevity of engineering components. These phenomena, individually or in combination, can have a substantial economic impact on industries as they necessitate repairs, maintenance, or even the replacement of engineering parts. The consequences are particularly critical in sectors such as petrochemical complexes or power plants, where the failure of a single component can have far-reaching consequences. In response, researchers have focused on mitigating or eliminating the adverse effects of these phenomena on the operational lifespan of engineering parts. Approaches include design modifications, material selection, lubrication techniques, and alterations to environmental conditions.

Cladding processes have emerged as a crucial means of reducing the destructive impacts of wear, corrosion, and fatigue. Cladding can be applied through various methods, including surface modification, alloying, and traditional coating techniques. Among these, surface cladding plays a vital role in engineering applications, as a significant portion of failures resulting from wear, corrosion, or fatigue originate from the surface. By employing wear and corrosion-resistant coatings, industrial components can be protected, ensuring their performance remains intact even under harsh environmental conditions. This approach alters the surface and internal properties of the parts to align with specific working environments and requirements [1].

Cladding methods for parts can be categorized based on various factors, such as cost, size, and efficiency. From a conventional perspective, the deposition process in coating can be classified into four main categories: gas deposition, solution deposition, melt deposition, and solid deposition.

By understanding and employing different coating techniques, researchers and industry professionals can address the challenges posed by wear, corrosion, and fatigue, ultimately extending the useful life of engineering components and enhancing the overall efficiency and reliability of industrial systems.

Cladding methods can be further classified based on the thickness of the coating applied. Various techniques fall under the umbrella of low-thickness coatings, typically measuring less than a micrometer. These methods include ion implantation, ion cladding, electrolytic and electroless plating, and deposition through chemical and physical vapor methods. In contrast, thick cladding methods involve

DOI: 10.1201/9781003492191-6

cladding ranging from several thousand micrometers to several millimeters in thickness. Examples of these techniques include thermal spraying, cold spraying, friction coating, electrochemical plating, and layer welding [2].

This classification plays a vital role, as cladding thickness significantly impacts coating efficiency. For instance, when applying thin coatings, mechanical stress becomes a critical consideration. In such cases, a thicker coating is desirable, as it enhances the part's ability to withstand stress and wear. By selecting an appropriate cladding thickness based on the specific requirements and operational conditions, researchers and practitioners can optimize the performance and durability of coated components. Investigation of the distinction between low-thickness and thick cladding methods empowers practitioners to make informed decisions when selecting the most suitable technique for their applications. By considering the coating thickness in relation to the desired performance characteristics, the efficiency and effectiveness of the cladding process can be enhanced.

The laser cladding process stands out as one of the most prominent cladding methods, with the resulting cladding thickness ranging from a few tenths of a millimeter to a few millimeters. Laser cladding has gained significant attention across various industries [3]. Market forecasts indicate substantial growth for the laser coating market in Europe from 2020 to 2024, with a projected increase of 41%. Laser cladding offers an attractive solution, enabling the repair of sensitive parts that were previously difficult to restore.

To provide a comprehensive understanding, the laser cladding process will be compared to other cladding methods, such as multilayer welding techniques (GMAW, GTAW, and SMAW) and thermal spraying (specifically, high velocity oxygen fuel spraying, or HVOF), based on process properties and conditions. This comparison aims to highlight the distinctive features and advantages of laser cladding. A detailed overview of this comparison is provided in the study [4].

By examining the unique characteristics of the laser cladding process in contrast to alternative methods, researchers and industry professionals can make informed decisions about the most suitable cladding technique for their specific requirements. Laser cladding's capabilities in repairing sensitive components, coupled with its versatility across industries, make it a compelling choice in the realm of cladding technologies.

The analysis study [4] reveals significant advantages of the laser cladding process compared to GTAW, SMAW, and HVOF processes. Firstly, the layering speed of laser coating is notably higher than that of GTAW, SMAW, and HVOF. This allows for quicker production and enhanced productivity in the laser cladding process. Additionally, laser coating induces less distortion in the workpiece compared to GMAW, GTAW, and SMAW processes, resulting in improved dimensional stability and reduced post-processing requirements.

Considering dilution, both laser coating processes and HVOF demonstrate superior performance compared to other methods. Dilution refers to the volume of the coating's surface area relative to its total volume and is crucial for ensuring a complete connection between the coating and the workpiece. It is recommended to maintain a dilution rate of approximately 2%–5% to achieve optimal bonding between the coating and the substrate [5].

Given the significance of these findings, it becomes essential to provide a simplified understanding of the laser cladding process, enabling readers to grasp the fundamentals of this technology more comprehensively. By presenting the details of the laser coating process in clear and accessible terms, this research aims to bridge the gap between complex technical concepts and practical applications, foster greater knowledge dissemination, and facilitate the adoption of laser cladding technology.

6.2 UNVEILING THE LASER CLADDING PROCESS: SIMPLIFIED INSIGHTS AND EXPLANATION

In the realm of laser material processing, the laser cladding process takes center stage when the objective is to create cladding or coating. Laser cladding involves a precise melting operation where a laser beam illuminates the workpiece, resulting in the formation of a protective layer [6]. This layer is generated by the interaction between the laser beam and a secondary material, commonly referred to as the "additive material" [4]. The additive material can be introduced to the workpiece through various methods, either during or prior to laser irradiation. Study [7] provides a schematic representation of this process.

In laser cladding, the focused laser beam acts as a catalyst, initiating the melting of the additive material while interacting with the workpiece. Through this controlled fusion, a layer of the additive material is deposited onto the surface of the workpiece, enhancing its properties and performance. This process offers a versatile solution for various applications, enabling the creation of protective coatings with improved wear resistance, corrosion resistance, and other desirable characteristics. The injection of the additive material can be accomplished using diverse techniques, allowing flexibility in the process. These techniques can range from direct injection during laser irradiation to pre-coating the workpiece prior to laser treatment. Each approach offers unique advantages and considerations that are tailored to meet specific requirements and desired outcomes.

The laser cladding process represents a form of direct energy deposition technique wherein the added material is melted and fused onto the surface of the workpiece. In this process, the powder is introduced into the nozzle's powder channel from a dedicated powder supply source, accompanied by a carrier gas. As the laser beam interacts with the workpiece, it initiates the melting of both the workpiece and the powder jet, resulting in the formation of a cladding layer on the workpiece surface [8]. Different feeding methods categorize the laser coating process into three main divisions: (1) pre-placed, (2) wire, and (3) powder [9]. This section will explore and examine each of these processes individually, shedding light on their unique characteristics and functionalities.

6.2.1 PRE-PLACED LASER CLADDING METHODS

The pre-placed laser cladding process involves the utilization of a powder bed that encompasses the workpiece. The laser beam's interaction with the powder bed and the workpiece initiates the melting process, leading to the formation of a molten pool. Upon solidification, a cladding layer is created on the workpiece surface. However, it is important to note that this laser cladding method presents certain challenges, including difficulties in implementation, time consumption, and the tendency to produce porous coatings. Furthermore, achieving a consistent thickness across large dimensions proves challenging with this method. The pre-placed laser cladding process is suitable for creating single-layer claddings, primarily on flat beds, and may not be viable for workpieces with complex three-dimensional geometries [10].

The pre-placed laser cladding process involves meticulous preparation and execution to achieve precise results. A bed of powder, serving as the precursor material, is strategically positioned over the workpiece. The laser beam, carefully directed onto the powder bed and workpiece, induces localized melting, leading to the formation of a molten pool. As the molten pool solidifies, a cladding layer is generated, exhibiting distinct regions within the cladding sample.

1. **Cladding**: The cladding area represents the desired layer that imparts specific properties and characteristics to the workpiece's surface. This region plays a crucial role in enhancing wear resistance, corrosion resistance, or other desired attributes.
2. **Overlap or dilution**: The overlap or dilution region occurs at the boundary between adjacent cladding tracks. It results from the slight overlap or fusion of the molten pools, leading to the incorporation of both the precursor material and the underlying workpiece material. The extent of overlap or dilution impacts the properties and quality of the resulting cladding.
3. **Heat-affected zone (HAZ)**: The heat-affected zone refers to the portion of the workpiece surrounding the cladding area that experiences significant thermal exposure during the laser cladding process. This region undergoes thermal transformation, affecting the microstructure and properties of the workpiece material adjacent to the cladding.
4. **Workpiece**: The underlying workpiece represents the substrate material onto which the cladding is applied. Its properties and structural integrity play a crucial role in the overall performance of the coated component.

The pre-placed laser cladding process offers a unique approach to surface enhancement, involving the use of a powder bed to facilitate controlled material deposition. While it presents challenges related to implementation, time consumption, and the tendency to produce porous coatings, this method is suitable for creating single-layer claddings on flat beds. Understanding the distinct regions within each coated sample, including the cladding, overlap or dilution, heat-affected zone, and the underlying workpiece, provides valuable insights into the complexities of the pre-placed laser cladding process. Continued research and development efforts are essential to overcome its limitations and expand its applicability, enabling precise and reliable surface enhancements for various industrial applications.

In 2010, Fallah et al. conducted a comprehensive study [11] to explore the solidification and phase change behavior in the pre-embedded laser coating process. The focus was on depositing Ti45Nb titanium powder onto low-carbon steel. Experimental parameters included a laser power of 1,100 W, a beam diameter of 1.2 mm, and a laser scanning speed ranging from 900 to 3,300 mm/min. The power density reached 970 W/mm^2 with an interference time spanning from 0.022 to 0.080 seconds. Study [11] illustrates the results obtained from this investigation, showcasing the cross-section of the coating created at a laser power of 1,100 W and an interference time of 0.024 seconds in part A. Additionally, micrographs obtained via SEM imaging for the overlap and middle sections of the cladding are presented in the study [11].

Study [11] highlights a successful bond between the coating and the workpiece achieved through the pre-embedded laser coating process. Study [11] provides insight into the microstructure development under specific conditions, namely a very short interference time (ranging from 0.024 to 0.027 seconds) and a laser power of 1,100 W. In this case, the coating's microstructure exhibits plate-like growth originating from the overlap region, with fine dendritic phases extending in a directional manner. These phases display a uniform distribution within the middle section of the coating. Notably, the study observed that the interference time governs the solidification behavior of the process while the power density remains constant. For extremely low interference times (0.024–0.027 seconds), a crack- and porosity-free coating with a remarkable hardness exceeding 1,000 Rockwell was achieved. Conversely, longer interference times (0.030–0.048 seconds) led to the formation of brittle intermetallic phases.

6.2.2 Laser Cladding by Wire

The wire laser cladding process involves the injection of added material in the form of wire into the molten pool. The laser beam melts both the workpiece and the wire, resulting in the formation of a molten pool. Upon solidification, a coating is formed on the workpiece. This method offers advantages such as environmental cleanliness and the utilization of 100% of the added material.

In a 2017 study, the wire laser cladding process was employed to coat a workpiece made of 316L stainless steel with Inconel 625 wire having a diameter of 1.14 mm [12]. The experimental parameters included a working power of 2,600 W, a speed ranging from 0.3 to 0.42 m/min, a feeding rate between 1.4 and 2 m/min, and an input heat ranging from 371 to 520 J/mm. The maximum coating thickness and width were reported as 3.09 and 4.04 mm, respectively [12].

In the study conducted on the wire laser cladding process [12], an investigation of the cladding cross-section's width and height revealed a direct relationship between the input heat and the size of the molten pool and temperature. This study showcased the presence of cell grains, devoid of secondary dendrites, near the workpiece and the overlapping area between the coating and the workpiece. These relatively large cell grains, measuring around 10 μm, were observed in limited regions at the lower part of the coating. Due to the relatively high cooling and freezing rates in the laser coating process, these areas did not facilitate the formation of secondary dendrites. In contrast, in the regions above the overlap, where the temperature gradient and freezing rate were reduced, columnar dendrites developed within the primary dendritic structure. The primary dendrite arm spacing was measured to be 5.5 μm, while the secondary dendrite arm spacing was determined to be 3.6 μm. Transitioning to the upper and surrounding areas of the coating, the microstructure shifted from oriented dendrites to coaxial grains. In these regions, characterized by a lower temperature gradient, grains with an average size ranging from 5 to 8 μm were formed.

Furthermore, in a separate study conducted in 2021, the laser hot wire coating (LHWC) process was explored for its benefits, including a high deposition rate [13]. The LHWC process involved injecting a 1.2-mm-diameter 308-L stainless steel wire into the molten pool at a temperature close to its melting point. Achieving this was made possible by employing a resistance heating device. As a result, the wire feeding rate increased, enhancing layering efficiency. The research involved layering three coatings of the stainless steel wire onto a low carbon steel workpiece with a 10 mm thickness. Process parameters included a laser power of 5,500 W, a speed of 0.36 m/min, a wire angle of 45°, a wire injection speed of 12 m/min, and the use of argon as a shielding gas with a flow rate of 30 L/min [13].

Based on the findings presented in the studies, the wire laser cladding process demonstrated the formation of a coating with a thickness of 11.84 mm and a width of 11.45 mm. Moreover, the heat-affected zone (HAZ) was observed to have a depth of 1.58 mm, indicating a successful and seamless connection between the coating and the workpiece without the presence of cracks or porosity. To gain further insight into the microstructure of the coating, the study [13] provides microscopic images of different sections of the coating. These results highlight the effectiveness of the wire laser cladding process in achieving substantial coating thickness and width while ensuring a well-integrated HAZ with excellent integrity. The microscopic images offer valuable visual representations of the coating's microstructure, allowing for a detailed examination of its composition and quality. Such insights contribute to the ongoing advancements in wire laser cladding technology and its applications in various industries. Further research and analysis in this field will continue to refine the understanding and utilization of wire laser cladding processes for the development of high-quality cladding [13].

The cladding layer of 308 L in wire laser cladding generally exhibits a microstructure consisting of gamma austenite, with a small amount of delta ferrite phase present within the gamma austenite dendrites. This study [13] illustrates the connection between the workpiece and the coating. The melting line, located at the center, acts as a boundary between the coating and the heat-affected zone (HAZ). No porosity, cracks, or other defects were observed in the melting line. Within the HAZ, the ferrite grain size is significantly smaller than that of the original workpiece due to the recrystallization process. During the initiation of the laser cladding, the laser beam irradiates the workpiece, causing the hot wire to melt and form a molten pool. Heat is then conducted from the molten pool to the workpiece. This region experiences the highest temperature gradient, the slowest grain growth rate, and a very high crystallization rate. Consequently, small plate-like crystals form at the center of the first layer. As the solid/liquid layer moves, the temperature gradient decreases, leading to an increase in the grain growth rate and the formation of columnar dendrites.

Study [13] illustrates the microhardness measurements taken across three different layers of the cladding in wire laser cladding. The results indicate that the initial hardness of the workpiece is relatively low but gradually increases, reaching its maximum in the heat-affected zone (HAZ). This hardness increase can be attributed to the processes of recrystallization, grain size reduction, and overall increased hardness in the HAZ region. The hardness of the first cladding area is similar to that of the HAZ, but it subsequently decreases as the distance from the HAZ increases. However, this decrease is not a continuous trend. Initially, the hardness decreases, followed by a region where the hardness remains relatively constant due to the interference of two layers. Beyond this region, the hardness decreases once again. At the boundary between the layers, concentric grains are observed, which contribute to an increase in hardness. Conversely, in the central areas of each layer, coarse dendritic crystals are present, leading to a reduction in hardness.

6.2.3 Laser Cladding by Powder

In the laser cladding by powder process, the injection of powder in relation to the laser beam can be categorized into two main methods: coaxial and non-coaxial [3]. This study provides a schematic representation of these two approaches [3].

In the co-axial powder laser cladding process, the integration of powder injection and laser beam transmission occurs directly on the workpiece. One of the advantages of the coaxial powder system is its ability to function independently of the movement direction. However, laboratory results indicate that the powder efficiency, defined as the ratio of powder applied to the workpiece to the injected powder per unit time, is lower in the coaxial powder method compared to the off-axis or lateral powder laser cladding. To enhance efficiency in both methods, the powder can be preheated while passing through the nozzle [3].

Several factors influence the interaction between the powder and the molten pool, such as the type of nozzle, the angle of powder injection relative to the beam, and the diameter of the powder flow upon entering the molten pool. A suitable nozzle is one that minimizes the presence of solid (unmelted) powder, as reducing unmelted powder and surface promotes increased "Powder catchment efficiency" [3]. Aside from the aforementioned considerations, the powder injection system plays a crucial role in determining the quality of the resulting cladding. To investigate the impact of nozzle diameter on various properties, including powder flow diameter, powder feed rate stability, powder profile homogeneity, and powder particle velocity, experiments have been conducted with the goal of capturing the powder flow through photography. These tests have revealed the existence of a relationship, denoted as (6.1), that governs the flow of powder:

$$d = \lambda z^2 + d_0 \tag{6.1}$$

In this relationship, the diameter of the profile at a given distance z (in mm) is represented by d, the qualitative coefficient of the powder profile (in 1/mm) is denoted by λ, z refers to the distance from the nozzle tip (mm), and d_0 represents the diameter of the nozzle (mm). The laboratory results pertaining to these parameters are presented in the study [3]. Based on this relationship, further details regarding each of the two systems will be discussed in separate sections to provide a comprehensive understanding of their characteristics.

6.2.3.1 Off-axis or Lateral Laser Powder Cladding

When the powder is injected off-axis in relation to the laser beam on the workpiece, changes in the direction of workpiece movement result in the creation of a coating with varying local conditions. As a consequence, the resulting coating becomes dependent on the direction of workpiece movement, leading to increased process sensitivity [14].

In a 2020 study [15], the interference between the powder flow and laser beam during non-coaxial powder injection was investigated. An optical detection system was employed to examine the movement and heat effects on the flow of TiC powder particles ranging in size from 20 to 40 μm. The coating process utilized a CO_2 pulsed laser with a power of 1 kW, a frequency of 90 kHz, a powder injection angle of 45°, and a powder injection diameter of 1.5 mm with argon gas as the carrier medium. Study [15] provides an overview of the laboratory equipment, a diagram illustrating the laser head and powder setup, and an actual image depicting the powder jet and laser beam.

The mentioned study on laser lateral powder cladding presented several significant findings. Firstly, it was reported that the average speed of TiC particles in the transfer flow was significantly lower (4–9 times) compared to the gas speed. This disparity was attributed to the impact of the powder particles on the injection pipe wall, with the powder particles exhibiting movement speeds ranging from 32 to 43 m/s, while the gas flow speed ranged from 47 to 189 m/s. Secondly, the study observed that when the laser beam interacted with the powder jet, it not only elevated the temperature of the powder to temperatures exceeding 3,600 K but also caused the powder particles to accelerate in the direction of the laser radiation, thereby propelling them towards the molten pool. It was further noted that the influence of the laser beam on particle acceleration and direction increased with lower powder injection speeds. These findings shed light on the intricate dynamics involved in laser lateral powder cladding, highlighting the impact of powder particle-wall interactions and the role of the laser beam in both heating the powder and directing its movement towards the molten pool [15].

6.2.3.4 Coaxial Laser Powder Cladding

When the powder is injected coaxially with the laser beam on the workpiece, the laser cladding process becomes independent of the cladding direction. Consequently, different parts of the resulting cover can be formed independently based on the movement direction of the workpiece. This advantageous characteristic makes coaxial laser cladding suitable to produce three-dimensional parts.

Various studies have been conducted in the field of the coaxial powder laser coating process, shedding light on its applications and benefits. These studies provide valuable insights into the process and contribute to our understanding of coaxial laser cladding.

In a 2010 study focusing on laser powder cladding, researchers investigated the powder flow out of the nozzle during the coating process [16]. The findings of this study led to the classification of the nozzle output flow into three distinct regions: annular, transmission, and Gaussian. The criterion for this classification was based on the density of powder particles observed during the flow [16].

In another 2013 study, the behavior of powder in the laser coating process was explored using a 5 kW CO_2 laser machine on a workpiece composed of S235 stainless steel and Ti6Al4V titanium superalloy powder. The study encompassed a range of parameters, including a speed of 0.3–1.4 m/min and a feed rate of 30 g/min [17]. Observations from the study [17] revealed that the transfer behavior of the powder to the molten pool varied for different powder sizes, specifically 80, 40, and 20 μm, when employing the coaxial method.

In a recent study conducted in 2019, researchers focused on simulating and investigating the flow pattern and temperature distribution within the coaxial powder laser cladding process, supported by laboratory results [18]. Study [18] illustrates the presence of molten metal flow within the cross section of the cladding, with varying speeds across different regions. This fluid flow significantly impacts the resulting microstructure once the material solidifies. Within the molten pool, the velocity reaches a maximum value of 5×10^{-10} m/s. The absorbed energy induces surface tension and generates a high temperature gradient.

The highest speed within the center of the coating can be attributed to three primary factors. First, the upstream flow of hot metal is caused by convection heat transfer and buoyancy forces (indicated by the blue line). Second, the boundary flow of metal along the upper surface of the coating, known as the Marangoni line (depicted in pink). Lastly, the formation of a swirling melt flow (green line), originating from the top of the coating and circulating downwards within the melt pool due to changes in heat distribution. Overall, this study provides valuable insights into the flow behavior and temperature dynamics during coaxial powder laser cladding, shedding light on the complex interactions that occur within the molten pool and their influence on the resulting microstructure.

6.2.4 Comparison between Three Types of Laser Cladding Processes

In the previous sections, we have explored and reviewed various laser cladding processes. Now, the question arises: what are the advantages and disadvantages of each process compared to others? And under what conditions should each process be chosen? In other words, when it comes to repairing, renovating, or producing parts using laser cladding, which process should be selected? It is important to note that this is a broad question, and answering it requires a deep understanding of the specific problem, details, and conditions. To gain a better understanding of the advantages and disadvantages of each laser cladding process, let's examine them in more detail based on the conditions.

The prefabricated laser cladding process is one of the simplest methods for repairing and manufacturing parts. However, it does come with certain challenges. During the laser cladding process, the pre-installed protective gas can cause the powder in the substrate to scatter. To mitigate this issue, the powder is often mixed with a chemical adhesive solution, which provides cohesion to the powder bed during the process. It is important to note that the chemical adhesive solution tends to evaporate during the process, resulting in porosity in the cladding. Additionally, the high dilution created in the workpiece during the pre-embedded laser cladding process usually limits it to a single layer [3].

The choice of laser cladding process directly affects the temperature and distortion of the workpiece. Studies indicate that although both powder and wire laser cladding processes yield desirable properties, the powder coating, for the same power, speed, or heat input, tends to have almost twice the thickness of wire cladding. Furthermore, due to the higher temperatures generated in the workpiece during the powder laser cladding process, it tends to induce more significant distortion in the workpiece [6].

In a study conducted in 2016, the laser coating process of Inconel 625 on a workpiece made of the same material was carried out using powder (size: 44–149 μm) and wire (diameter: 0.76 mm) forms, with the same power of 2.5 kW, travel speed of 10.6 mm/s, and input heat of 236 J/mm [6]. Study [6] demonstrates that for the same input heat, the powder coating thickness is approximately 2.6 times that of the wire coating. Moreover, this study shows that the measured temperature in the region of the workpiece subjected to the powder process is higher than that of the wire process. Consequently, it can be expected that the distortion resulting from the powder process is more pronounced compared to the wire process. It is important to mention that the distortion and temperature were measured using a Keyance LK-031 laser displacement sensor and an Omega GG-K-30 thermocouple, respectively.

In another study conducted in 2017, the laser coating of Inconel 625 on 304 stainless steel workpieces was carried out using both powder and wire methods, and the results were analyzed. Powder with a size range of 40–260 μm and wire with a diameter of 1.2 mm were employed. The processes were performed at the same laser power of 1.8 kW and laser scanning speed of 100 mm/min. The findings of this study revealed that the average coating thickness obtained from the powder process was 3.2 mm, while the wire process resulted in an average coating thickness of 2.7 mm. Additionally, the coating width measured 3.4 mm for the powder coating and 4.7 mm for the wire coating. Microscopic imaging of the resulting coatings from both the wire and powder laser processes can be observed [19]. The microscopic images provide insights into the formation of dendrites in both the powder and wire laser coating processes, highlighting a notable distinction. In the powder laser coating process, smaller coaxial dendrite grains are formed compared to the wire laser coating process. In other words, the wire process predominantly yields vertically growing columnar dendrites extending from the coating-base metal interface, while the powder process results in finer grains due to the higher cooling rate.

Study [19] presents the microhardness of the coatings produced by the wire and powder processes, respectively. These results demonstrate that the powder process yields a coating hardness of 245 Vickers, while the wire process produces a hardness of 224 Vickers. This difference can be attributed to the presence of secondary deposits (hard phase) in the interdendritic regions and the grain size within the dendritic core. An increase in the number of deposits or a reduction in grain size leads to higher microhardness values [19]. Considering these findings, it becomes evident that the various types of laser coating processes can be summarized and compared based on previous studies. To highlight their significance, Table 6.1 provides a comparison of the advantages, disadvantages, and applications of the three methods—prefabricated, wire, and powder.

The laser cladding process offers various methods for surface hardening and covering surfaces, each suited for different applications. Table 6.1 provides an overview of these methods and their respective advantages and limitations. Pre-placed laser cladding is a commonly employed technique for surface hardening. This process offers high efficiency, making it an ideal choice for applications where hardening surfaces is the primary objective. However, its usage has declined in recent years due to challenges related to high sensitivity and the difficulty of adding paste. The wire method is specifically employed for covering large surfaces and relatively simple geometries. It offers good coverage but has limitations when it comes to cladding complex geometries. Therefore, it may not be the most suitable option for intricate shapes or components with complex designs.

For complex geometries and prototyping, powder laser cladding is often preferred. This method provides high precision and the ability to create a uniform surface. However, it should be noted that powder laser cladding has some disadvantages, such as lower material efficiency and a tendency to create porosity within the cladding. To summarize, the selection of a laser cladding method depends on the specific requirements of the application. Pre-placed laser cladding offers high efficiency but is less commonly used due to challenges associated

TABLE 6.1
Comparison of Different Types of Laser Cladding Methods based on Advantages, Disadvantages, and Application

Method	Application	Advantages	Disadvantages	References
Pre-placed	Hardening surfaces	High material efficiency	High sensitivity, Difficulty adding paste	[3,20]
Wire	Large product, simple geometry	High material efficiency	Unusable in complex geometries	[8,21]
Powder	Complex geometry, prototype	High accuracy, providing uniform surface	Low material efficiency, tendency to porosity	[3,21]

with sensitivity and paste application. The wire method is suitable for covering large surfaces and simpler geometries, while powder laser cladding is preferred for complex geometries and prototyping, despite its drawbacks of lower material efficiency and potential porosity [22,23].

6.3 THE EFFECT OF PARAMETERS ON THE CHARACTERISTICS OF THE CLAD

The laser cladding process, like other laser material processing methods, is influenced by various variables. These variables have a direct impact on the solidification process, cladding formation, and ultimately, the characteristics of the provided clad [4]. Geometric characteristics, such as the width and height of the cladding, and mechanical characteristics like hardness, brittleness, yield stress, and fatigue, are important considerations when evaluating the provided clad. The parameters of the laser cladding process can be broadly categorized into two main groups: primary and secondary variables. In the upcoming sections, we will discuss each variable in detail and examine its specific effects on the characteristics of the provided clad. By investigating and manipulating these variables, it is possible to optimize the laser cladding process to achieve the desired clad characteristics. Fine-tuning parameters can help control clad dimensions, mechanical characteristics, and other essential features. This knowledge is crucial for ensuring consistent and reliable results in laser cladding applications.

6.3.1 Primary Processing Parameters

In this section, the impact of the primary process parameters on various clad characteristics, including hardness, porosity, dilution, and melt pool temperature, will be explored. Each parameter will be discussed individually to investigate its influence on these clad characteristics.

6.3.1.1 How Does Increasing the Laser Power Lead to a Decrease in Clad Hardness?

Laser power is a critical variable that significantly impacts the laser coating process. It refers to the amount of energy delivered by the laser beam to the workpiece within a given time period [24]. The power of the laser directly influences various characteristics of the provided clad.

A study focused on investigating the effect of laser power on the laser-clad process of NiCrBSi powder on a 42CrMo steel workpiece [25]. This study illustrates the hardness diagram with respect to the coating height for laser powers ranging from 1,500 to 3,500 W. The figure provides insights into the relationship between laser power and the clad hardness at different clad heights. By analyzing this diagram, it becomes possible to understand the optimal power range for achieving the desired hardness of the clad. It is worth noting that the specific results of the study were not provided in the given context. The findings of such studies help establish guidelines for selecting an appropriate laser power based on the desired clad characteristics. Investigating the influence of laser power on the coating process enables researchers and practitioners to make informed decisions and optimize the laser cladding process for specific applications.

As depicted in the study [25], the microhardness of the coating decreases as the laser power increases. This trend can be attributed to the increased influx of iron into the clad, which subsequently reduces its hardness [25]. Similar findings have been observed in other studies as well, such as those involving H13 tool steel powder laser cladding on the same material [26] and TiB_2-TiC-Al_2O_3 cladding on 1020 steel [27]. The increase in power during the laser cladding process promotes greater melting and mixing of the base material with the cladding material. This results in a higher iron content in the cladding, which tends to lower its hardness. It is important to note that the specific experimental conditions, materials, and measurement techniques used in these studies may vary, but the overall trend of decreasing microhardness with increasing laser power remains consistent.

These consistent findings emphasize the importance of considering the effect of laser power on the clad characteristics. Achieving the desired hardness requires careful control and optimization of the power parameter, considering the specific materials and application requirements. By investigating the relationship between laser power and clad hardness, researchers and practitioners can make informed decisions to achieve optimal clad quality.

6.3.1.2 How Does the Diameter of the Laser Beam Affect the Dilution of the Clad?

The distance between the nozzle and the workpiece can impact the diameter of the laser beam during the laser cladding process. This parameter is often referred to as the focal length, which determines the focus and divergence of the laser beam. In a particular study, the impact of focal length was examined in three different modes: negative, focused, and positive decentralized, during the laser cladding process of SiC powder on magnesium alloy ZE41. By investigating the effect of these focal length modes, researchers sought to understand how variations in the distance between the nozzle and the workpiece affect the deposited clad [28].

Investigating the influence of focal length on the laser cladding process is crucial for achieving the desired deposited clad characteristics. Adjusting the focal length can have implications for the heat distribution, energy density, and interaction between the laser beam and the powder material, ultimately impacting the coating quality and characteristics.

Study [28] demonstrates a positive eccentric beam configuration where the smallest beam diameter is positioned below and inside the workpiece. On the other hand, this study represents the two negative modes: focused and defocused beams, respectively. In the negative focused mode, the smallest beam diameter lies on the surface of the workpiece, while in the negative defocused mode, it is above and outside the workpiece. Additionally, the figure presents the results of X-ray energy diffraction spectroscopy analysis (EDX) for each of the three beam modes [28]. These results provide valuable insights into the chemical composition of the coatings obtained in each configuration. When using a negative defocused beam, the powder is not entirely melted as the high-energy areas of the laser beam fail to reach it. Consequently, this limits the formation of a complete coating, as the predominant energy mechanism is focused on melting the workpiece itself. In the focused mode, which is considered the optimal configuration, the laser beam's energy is efficiently directed toward the surface of the powdered workpiece, leading to surface melting and favorable bonding between the coating and the workpiece. On the other hand, the positive eccentricity mode demonstrates increased melt pool size and dilution, which are undesirable effects. The EDX results further support and confirm these observations [28,29]. These findings underscore the significance of beam configuration in the laser cladding process. Choosing the appropriate mode enables the optimization of energy utilization, ensuring effective coating formation and desired bonding with the workpiece.

6.3.1.3 How Does Increasing the Advance Speed Lead to an Increase in Clad Hardness?

The laser beam advance speed refers to the speed at which the laser head moves during the laser-cladding process. It essentially determines the overall speed of the coating process, with higher advance speeds resulting in shorter processing times. In a particular study, researchers examined the influence of laser beam advance speed on the hardness of the coating obtained from P25 powder, which consists of 25% chromium and 10% nickel by mass, applied to carbon steel [30]. Study [30] illustrates the findings of this study. The figure likely presents a graphical representation showcasing the relationship between the laser beam advance speed and the resulting coating hardness. However, specific details or trends from the figure are not provided in the given context. By investigating the effect of the laser beam advance speed on coating hardness, researchers aimed to understand the optimal processing conditions for achieving the desired hardness level. The results of this study, not specified here, provide insights into the relationship between advance speed and coating hardness, assisting in the optimization of the laser cladding process for P25 powder on carbon steel. Further analysis and interpretation of the study's results would be necessary to extract specific conclusions regarding the impact of laser beam advance speed on coating hardness.

Based on a study [27], it was observed that an increase in the laser beam advance speed leads to higher surface hardness. This trend can be attributed to the corresponding increase in the cooling rate as the speed rises. The accelerated cooling rate promotes the formation of a finer grain structure in the coating. Consequently, it can be predicted that the hardness will increase with higher advance speeds. The relationship between advance speed and surface hardness can be explained by the microstructural changes that occur during rapid cooling. When the cooling rate is higher, the molten material solidifies more rapidly, resulting in smaller and more closely packed grains. This finer grain structure is associated with increased hardness due to the increased number of grain boundaries, which can impede dislocation movement and enhance mechanical characteristics. However, it is important to note that there may be an upper limit beyond which further increasing the advance speed could have diminishing returns or adverse effects on cladding quality. The specific threshold and the optimum advance speed depend on various factors, including the material being clad, powder composition, and processing conditions. The findings from this study provide valuable insights into the relationship between laser beam advance speed and clad hardness. These insights can guide the selection of appropriate processing parameters to achieve the desired clad characteristics for P25 powder on carbon steel.

6.3.1.4 How Does the Feed Rate Lead to Increased Porosity in the Clad?

The speed at which secondary materials enter the molten pool in the laser cladding process is referred to as the feed rate. In the case of powder coating, this quantity can be equated to the output powder flow rate and is typically measured in grams per minute. The impact of feed rate on the porosity of the coating, both in terms of height and width, was investigated in a study focusing on the laser coating process using Inconel 718 for both the powder and workpiece materials. The study was conducted at a laser power of 3,858 W and a speed of 1,500 mm/min. The findings of this study were summarized in study [31].

Study [31] indicates that an increase in the feed rate corresponds to an increase in porosity. The rationale behind this observation is that as the secondary material (powder) content increases, there is a higher likelihood that the laser energy may not be sufficient to fully melt the material. Consequently, the molten pool may not have enough time to form and expand adequately. Furthermore, as the feeding rate increases, it also results in an increased carrier gas flow since the secondary material is injected into the pool alongside the carrier gas. When the molten pool fails to sufficiently melt the powder, the excess carrier gas enters and occupies the void spaces. This, in turn, leads to an increase in porosity within the resulting coating [31]. The results from this study highlight the relationship between feed rate and clad porosity. Investigating this relationship helps in optimizing the feed rate parameter to minimize porosity and achieve clad with improved structural integrity and desired characteristics.

6.3.1.5 How Does Shielding Gas Affect the Temperature of the Molten Pool?

One common characteristic of laser material processing is the application of high energy within a short time, resulting in an elevated temperature of the workpiece. At high temperatures, metals are prone to oxidation. To prevent oxidation and maintain the quality of the coating, shielding gas is employed, as considered in studies [4,32].

In a specific study, the influence of different types of shielding gas on the temperature of the molten pool was investigated during the laser powder cladding process using a nickel-based powder on an Inconel 718 workpiece. The research findings were reported under the same laser power conditions of 400 W, using three different shielding gas compositions: pure argon, 75% argon-25% helium, and 50% argon-50% helium, as outlined in the study [33]. It is worth noting that argon and helium are commonly used shielding gases in the laser cladding process, and they were considered in this study. Furthermore, these gases do not participate in the chemical reactions of the process and do not impact its effectiveness [33].

By varying the shielding gas composition, the study aimed to determine its impact on the temperature of the molten pool during the cladding process. The results provide valuable insights into optimizing the choice of shielding gas to control the temperature and minimize oxidation. The selection of an appropriate shielding gas helps maintain the integrity of the cladding and ensures desired cladding characteristics, such as adhesion and structural stability. It is important to consider that the choice of shielding gas can vary depending on factors such as the material being cladded, process requirements, and desired outcomes. Therefore, investigating the effects of different shielding gases allows for informed decision-making and optimization of the laser cladding process.

Considering the study [33], an increase in the helium content from zero to 50% by volume leads to a rise in the average temperature of the molten pool, from approximately 1,938 to 2,097 K. This temperature increase can be attributed to the significantly higher heat transfer coefficient of helium compared to argon. In fact, helium possesses a heat transfer coefficient approximately 40 times higher than that of argon [34,35]. It is worth noting that as the temperature of the molten pool increases, the porosity of the resulting clad tends to decrease. Therefore, the type of shielding gas indirectly affects the porosity of the clad as well [33]. The higher heat transfer coefficient of helium allows for more efficient heat dissipation from the molten pool, resulting in a higher average temperature. This increase in temperature promotes better melting and fusion of the powder material, leading to reduced porosity in the clad. Consequently, by employing a shielding gas composition with a higher helium content, the clad quality can be improved through reduced porosity. The findings from this study highlight the importance of selecting the appropriate shielding gas composition in the laser cladding process. By investigating the impact of different gases on the temperature of the molten pool and, consequently, the porosity of the clad, researchers and practitioners can optimize the process parameters to achieve clad with desired characteristics and minimal defects.

6.3.2 Secondary Processing Parameters

In this section, we will examine the impact of sub-parameters on various properties in the laser coating process. These sub-parameters include overlap, direction of paths, pre-heating, and post-heating. Each sub-parameter will be discussed in separate subsections to carefully investigate their effects on specific coating properties.

6.3.2.1 How will the Overlap and Direction of the Paths Affect the Smoothness of the Surface and the Strength of the Clad?

To achieve a complete and uniform coating, it is common practice to position the clad paths next to each other with overlap, as depicted in the study [36]. The overlapping of paths serves two important purposes: preventing crack formation and ensuring a continuous connection between adjacent clad. During the laser cladding process, as a new path is laid down, it partially melts the previous path. This melting of the previous path allows it to participate in the molten pool of the new path, facilitating a smooth transition and bonding between adjacent clad. The proportion of the previous path that melts and contributes to the molten pool of the new path is referred to as "Overlap".

Study [36] provides a schematic representation of several clad paths, highlighting the overlay height and width as well as the concept of overlap. The overlap helps in maintaining clad integrity, preventing discontinuities and potential cracks between adjacent paths. By strategically controlling the overlap, the coating process can ensure the formation of a cohesive and uniform clad layer. The inclusion of overlapping paths, as shown in the study [36], is an essential technique employed in the laser cladding process to achieve consistent and reliable claddings. Careful consideration of overlay height, width, and overlap is necessary to optimize the cladding quality and ensure proper bonding between successive paths.

In a study, the combined effect of power and overlap on surface smoothness was investigated. It was found that at low powers (1,200 W), a low overlap resulted in increased distance between coating paths, leading to weaker and poorer connections between the secondary material and the workpiece. Consequently, this decreased surface smoothness and adversely affected the coating quality. On the other hand, at high powers (1,600 W), increased overlap enhanced the contact between overlap points and the workpiece. This resulted in higher surface smoothness ratios and improved coating quality [37]. It is important to note that the objective of the study was to achieve minimal surface smoothness to eliminate the need for post-coating machining processes. Additionally, if the goal is to maximize the overlay height, studies suggest an overlap of 29.3% [38].

Apart from the overlap percentage, the direction of cladding also impacts the mechanical characteristics of the coated piece. In a different study, the mechanical strength of three clad directions (longitudinal, oblique, and transverse) for M300 steel powder and workpieces were compared with the uncoated state. Study [39] demonstrates the relationship between the direction and angle of the paths (α) and the direction of tension considered in the tension test. The three clad directions resulted in different mechanical characteristics. This study indicates that the clad direction influences both yield stress and Young's modulus parameters [39].

These findings highlight the significance of both overlap and coating direction in the laser cladding process. Optimizing these parameters allows for achieving the desired surface smoothness, cladding quality, and specific mechanical characteristics. Proper consideration of overlap and clad direction facilitates customization of the cladding process based on the intended objectives and requirements.

6.3.2.2 How Does Preheating Increase the Maximum Feeding of the Clad?

In the laser cladding process, preheating refers to the practice of heating the workpiece, the secondary material, or both before initiating the cladding process.

The primary objective of preheating is to reduce the likelihood of crack formation in the cladding. By gradually heating the workpiece, preheating helps to slow down the heating speed during the cladding process. This controlled heating process minimizes thermal gradients and thermal stresses, which in turn reduces the risk of cracks in the cladding. Preheating promotes better material flow, enhances the bond between the coating and the substrate, and improves the overall quality and integrity of the cladding [4].

In addition to crack prevention, some studies suggest that preheating in the laser cladding process can have other benefits. For instance, preheating has been associated with increased energy efficiency in the process. By raising the initial temperature of the workpiece, less laser energy is required to achieve the desired coating temperature. This energy efficiency can result in cost savings and improved process productivity [40]. Furthermore, preheating has also been linked to an increase in the maximum injection feeding rate. The elevated temperature of the workpiece or secondary material helps to reduce their viscosity, enabling higher feed rates during the cladding process. This increased feeding rate can lead to improved cladding deposition rates and efficiency [40]. Overall, preheating in the laser cladding process plays a crucial role in reducing crack formation, enhancing energy efficiency, and potentially increasing the maximum feeding rate. It is an important technique for achieving high-quality claddings and optimizing the efficiency of the process.

In a study focused on the laser cladding process of NiCrBSi powder on a 42CrMo steel workpiece, the maximum injection feeding rate (φ_{max}) and maximum energy efficiency (Ψmax) were investigated. The study aimed to understand the impact of preheat temperature on these parameters. Study [25] presents the results of this investigation. According to the findings, increasing the preheat temperature correlates with an increase in both the maximum injection feeding rate and energy efficiency [25]. The reason behind this relationship lies in the distribution of laser energy during the coating process. As the preheat temperature increases, less energy from the laser is required to heat the workpiece up to its melting point. Consequently, a larger portion of the laser energy can be utilized for the formation and maintenance of the molten pool. This higher utilization of laser energy for the molten pool formation leads to an increase in energy efficiency. Moreover, with more laser energy focused on the molten pool, there is greater capacity for injecting and incorporating secondary material into the process. Consequently, the maximum injection feeding rate of the laser cladding process increases. This allows for a higher rate of material deposition and can potentially improve the efficiency of the cladding process. The results of this study emphasize the positive effects of preheat temperature on both the maximum injection feeding rate and energy efficiency in the laser cladding process. By optimizing the preheat temperature, researchers and practitioners can enhance process performance, achieve higher deposition rates, and improve the overall efficiency of the laser cladding process.

6.3.2.3 How Does Post-Heat Affect the Hardness of the Deposited Clad?

In the laser cladding process, heating the clad part after the completion of the process is referred to as post-heating. The impact of post-heating on the hardness of the laser-clad coating was investigated in a study involving the laser cladding of 410 steel powder on a hypereutectoid steel workpiece. The study aimed to understand the effect of post-heating on the hardness profile of the resulting cladding. The results of this investigation were presented in this study, comparing the hardness profiles in the absence and presence of post-heating at a temperature of 350°C [41].

Study [41] likely showcases the hardness profile of the laser-clad coating, indicating variations in hardness across the depth of the coating. By comparing the profiles of the unheated (A) and post-heated (B) claddings, the influence of post-heating on the hardness distribution can be analyzed. It is expected that the post-heating process at 350°C will have an impact on the microstructure and subsequent hardness of the cladding. Post-heating can aid in stress relief, promote grain growth, and enhance the metallurgical bond between the cladding and the substrate. These effects may contribute to changes in the hardness profile of the cladding. A thorough examination and interpretation of the results presented in this study would be required to draw specific conclusions about the impact of post-heating on the hardness of the laser-clad coating in the studied system. The findings of this study contribute to understanding the potential benefits of post-heating in optimizing the characteristics and performance of laser-clad coatings.

According to this study, the presence of post-heating in the laser cladding process of 410 steel powder on a hypereutectoid steel workpiece resulted in a decrease in the average hardness of the clad. The average hardness values reported were 475 Vickers in the presence of post-heating and 517 Vickers in its absence. Consequently, the presence of post-heating led to a reduction in cladding hardness and, therefore, decreased brittleness [41].

In a separate study focusing on the laser cladding of NiCrBSi powder on 410 stainless steel, it was observed that increasing the preheat temperature caused the nickel grains within the cladding to grow, which in turn reduced the resistance to dislocation movement. As a result, the resistance to deformation caused by external forces decreased, leading to a reduction in surface hardness [42]. This finding was supported by another study examining the laser cladding process of 70Ni-30C powder on Ti6Al4V alloy [43]. These research results collectively suggest that increasing the preheat temperature or applying post-heating in the laser cladding process can contribute to the reduced hardness of the coating. The growth of grains and decreased resistance to dislocation movement are potential factors leading to this phenomenon. The observed reduction in cladding hardness can influence the clad's mechanical characteristics, such as its resistance to deformation and brittleness. The findings highlight the importance of considering the effects of preheat temperature and post-heating in the laser cladding process and their implications on cladding hardness and related characteristics. Further investigation and analysis are necessary to investigate the underlying mechanisms and optimize the process parameters to achieve the desired clad characteristics.

REFERENCES

[1] S. Zanzarin, *Laser Cladding with Metallic Powders*. University of Trento, 2015.

[2] A. Matthews and D. S. Rickerby, *Advanced Surface Coatings: A Handbook of Surface Engineering*. Springer, 1991.

[3] J. R. Davis, *Surface Engineering for Corrosion and Wear Resistance*. ASM International, 2001.

[4] E. Toyserkani, A. Khajepour, and S. F. Corbin, *Laser Cladding*. CRC Press, 2004.

[5] R. Colaco, L. Costa, R. Guerra, and R. Vilar, "A simple correlation between the geometry of laser cladding tracks and the process parameters," *Laser Processing: Surface Treatment and Film Deposition,* pp. 421–429, 1996.

[6] J. Ion, *Laser Processing of Engineering Materials: Principles, Procedure and Industrial Application*. Elsevier, 2005.

[7] J. Heigel, M. Gouge, P. Michaleris, and T. Palmer, "Selection of powder or wire feedstock material for the laser cladding of Inconel(r) 625," *Journal of Materials Processing Technology,* vol. 231, pp. 357–365, 2016.

[8] M. Schmidt, P. Huke, C. Gerhard, and K. Partes, "In-line observation of laser cladding processes via atomic emission spectroscopy," *Materials,* vol. 14, no. 16, p. 4401, 2021.

[9] R. Vilar, "Laser cladding," *Journal of Laser Applications,* vol. 11, no. 2, pp. 64–79, 1999.

[10] L. Zhu *et al.*, "Recent research and development status of laser cladding: A review," *Optics & Laser Technology,* vol. 138, p. 106915, 2021.

[11] D. S. Salehi, *Sensing and Control of Nd:YAG Laser Cladding Process*. Swinburne University of Technology, Industrial Research Institute Swinburne, 2005.

[12] V. Fallah, S. F. Corbin, and A. Khajepour, "Solidification behaviour and phase formation during pre-placed laser cladding of Ti45Nb on mild steel," *Surface and Coatings Technology,* vol. 204, no. 15, pp. 2400–2409, 2010.

[13] X. Xu *et al.*, "Research on microstructures and properties of Inconel 625 coatings obtained by laser cladding with wire," *Journal of alloys and Compounds,* vol. 715, pp. 362–373, 2017.

[14] W. Li *et al.*, "Microstructure evolution and mechanical properties of 308L stainless steel coatings fabricated by laser hot wire cladding," *Materials Science and Engineering: A*, vol. 824, p. 141825, 2021.

[15] U. De Oliveira, V. Ocelik, and J. T. M. De Hosson, "Analysis of coaxial laser cladding processing conditions," *Surface and Coatings Technology,* vol. 197, no. 2–3, pp. 127–136, 2005.

[16] I. Gulyaev, O. Kovalev, P. Pinaev, and G. Grachev, "Optical diagnostics of radiation interaction with the powder stream laterally transported during laser cladding," *Optics and Lasers in Engineering,* vol. 126, p. 105877, 2020.

[17] I. Tabernero, A. Lamikiz, E. Ukar, L. L. De Lacalle, C. Angulo, and G. Urbikain, "Numerical simulation and experimental validation of powder flux distribution in coaxial laser cladding," *Journal of Materials Processing Technology,* vol. 210, no. 15, pp. 2125–2134, 2010.

[18] I. Smurov, M. Doubenskaia, and A. Zaitsev, "Comprehensive analysis of laser cladding by means of optical diagnostics and numerical simulation," *Surface and Coatings Technology,* vol. 220, pp. 112–121, 2013.
[19] B. Khamidullin, I. Tsivilskiy, A. Gorunov, and A. K. Gilmutdinov, "Modeling of the effect of powder parameters on laser cladding using coaxial nozzle," *Surface and Coatings Technology,* vol. 364, pp. 430–443, 2019.
[20] T. Abioye, P. Farayibi, and A. Clare, "A comparative study of Inconel 625 laser cladding by wire and powder feedstock," *Materials and Manufacturing Processes,* vol. 32, no. 14, pp. 1653–1659, 2017.
[21] E. Lugscheider, H. Bolender, and H. Krappitz, "Laser cladding of paste bound hardfacing alloys," *Surface Engineering,* vol. 7, no. 4, pp. 341–344, 1991.
[22] T. Abioye, D. McCartney, and A. Clare, "Laser cladding of Inconel 625 wire for corrosion protection," *Journal of Materials Processing Technology,* vol. 217, pp. 232–240, 2015.
[23] B. Borges, L. Quintino, R. M. Miranda, and P. Carr, "Imperfections in laser clading with powder and wire fillers," *The International Journal of Advanced Manufacturing Technology,* vol. 50, pp. 175–183, 2010.
[24] J. Nurminen, J. Riihimäki, J. Näkki, and P. Vuoristo, "Comparison of laser cladding with powder and hot and cold wire techniques," In: *Pacific International Conference on Applications of Lasers and Optics.* AIP Publishing, *2006.*
[25] W. M. Steen and J. Mazumder, *Laser Material Processing.* Springer, 2010.
[26] W. Kai-ming, F. Han-guang, L. Yu-long, L. Yong-ping, W. Shi-zhong, and S. Zhen-qing, "Effect of power on microstructure and properties of laser cladding NiCrBSi composite coating," *Transactions of the IMF,* vol. 95, no. 6, pp. 328–336, 2017.
[27] G. Telasang, J. D. Majumdar, G. Padmanabham, M. Tak, and I. Manna, "Effect of laser parameters on microstructure and hardness of laser clad and tempered AISI H13 tool steel," *Surface and Coatings Technology,* vol. 258, pp. 1108–1118, 2014.
[28] M. Masanta, S. Shariff, and A. R. Choudhury, "Evaluation of modulus of elasticity, nano-hardness and fracture toughness of TiB2-TiC-Al2O3 composite coating developed by SHS and laser cladding," *Materials Science and Engineering: A,* vol. 528, no. 16–17, pp. 5327–5335, 2011.
[29] A. Riquelme, P. Rodrigo, M. D. Escalera-Rodríguez, and J. Rams, "Analysis and optimization of process parameters in Al-SiCp laser cladding," *Optics and Lasers in Engineering,* vol. 78, pp. 165–173, 2016.
[30] J. Liu, H. Yu, C. Chen, F. Weng, and J. Dai, "Research and development status of laser cladding on magnesium alloys: A review," *Optics and Lasers in Engineering,* vol. 93, pp. 195–210, 2017.
[31] S. Barnes, N. Timms, B. Bryden, and I. Pashby, "High power diode laser cladding," *Journal of Materials Processing Technology,* vol. 138, no. 1–3, pp. 411–416, 2003.
[32] C. Zhong, T. Biermann, A. Gasser, and R. Poprawe, "Experimental study of effects of main process parameters on porosity, track geometry, deposition rate, and powder efficiency for high deposition rate laser metal deposition," *Journal of Laser Applications,* vol. 27, no. 4, 2015.
[33] B. Graf, S. Ammer, A. Gumenyuk, and M. Rethmeier, "Design of experiments for laser metal deposition in maintenance, repair and overhaul applications," *Procedia CIrP,* vol. 11, pp. 245–248, 2013.
[34] J. E. Ruiz, M. Cortina, J. I. Arrizubieta, and A. Lamikiz, "Study of the influence of shielding gases on laser metal deposition of Inconel 718 superalloy," *Materials,* vol. 11, no. 8, p. 1388, 2018.

[35] J. Arrizubieta, I. Tabernero, J. E. Ruiz, A. Lamikiz, S. Martinez, and E. Ukar, "Continuous coaxial nozzle design for LMD based on numerical simulation," *Physics Procedia,* vol. 56, pp. 429–438, 2014.
[36] J. Elmer, J. Vaja, R. Pong, T. Gooch, and H. Barth, "The Effect of Ar and N2 shielding gas on laser weld porosity in steel, stainless steel, and nickel," *Welding Journal,* vol. 2015, no. LLNL-JRNL-663819, 2015.
[37] S. H. Oliari, A. S. C. M. D'Oliveira, and M. Schulz, "Additive manufacturing of H11 with wire-based laser metal deposition," *Soldagem & Inspeção,* vol. 22, pp. 466–479, 2017.
[38] R. R. Rashid *et al.*, "Effect of clad orientation on the mechanical properties of laser-clad repaired ultra-high strength 300M steel," *Surface and Coatings Technology,* vol. 380, p. 125090, 2019.
[39] G. Lian, M. Yao, Y. Zhang, and C. Chen, "Analysis and prediction on geometric characteristics of multi-track overlapping laser cladding," *The International Journal of Advanced Manufacturing Technology,* vol. 97, pp. 2397–2407, 2018.
[40] Y. Li and J. Ma, "Study on overlapping in the laser cladding process," *Surface and Coatings Technology,* vol. 90, no. 1–2, pp. 1–5, 1997.
[41] S. Zhou, Y. Huang, and X. Zeng, "A study of Ni-based WC composite coatings by laser induction hybrid rapid cladding with elliptical spot," *Applied Surface Science,* vol. 254, no. 10, pp. 3110–3119, 2008.
[42] T. Roy et al., "Effect of deposition material and heat treatment on wear and rolling contact fatigue of laser cladded rails," *Wear,* vol. 412, pp. 69–81, 2018.
[43] G. Durge, A. Chandak, A. K. Jaiswal, K. U. V. Kiran, B. R. Sunil, and R. Dumpala, "Effect of heat treatment on the hardness and wear characteristics of NiCrBSi laser clad deposited on AISI410 stainless steel," *Materials Research Express,* vol. 6, no. 8, p. 086524, 2019.

7 Laser Beam Cladding
Features and Applications

7.1 FEATURES OF LASER CLADDING

The laser cladding process, renowned for its versatility, unfolds several distinctive features that contribute to its effectiveness across a spectrum of industrial applications. This comprehensive investigation aims to dissect both the advantages and challenges inherent in laser cladding. On the advantageous front, its multifaceted benefits include heightened precision, diminished heat-affected zones (HAZs), and the capability to deposit a wide array of materials. These attributes render laser cladding particularly enticing for applications in the aerospace, automotive, and tooling industries.

However, navigating the landscape of laser cladding is not without hurdles. Challenges such as parameter optimization, material compatibility, and precise process control demand meticulous attention. Each of these challenges is explored in dedicated subsections to provide a nuanced understanding of their complexities. The intricacies of these aspects offer valuable insights, enabling a thorough comprehension of the potential benefits and obstacles tied to laser cladding.

This exploration serves as a vital knowledge base for informed decision-making in the adoption of this cutting-edge technology across diverse manufacturing sectors. Recognizing the interplay between advantages and challenges is paramount to harnessing the full potential of laser cladding and integrating it seamlessly into contemporary industrial practices. The intricate analysis presented here lays the groundwork for practitioners and researchers alike to navigate the dynamic landscape of laser cladding with acumen and foresight.

7.1.1 Advantages of Laser Cladding

With the rapid evolution of laser coating technology, recent research studies shed light on the key advantages of laser cladding. Central to its efficacy is the concept of "localize," where the laser beam's distinct characteristics—monochromaticity, directivity, and coherence—contribute to the emission of concentrated energy onto a small workpiece area. This results in a precise and controlled heat input, a critical factor in various industrial applications [1]. The high power density achieved through laser beam localization is a notable advantage [2]. Power density, defined as the amount of laser power radiated per unit area, distinguishes laser cladding from other industry cladding processes. For instance, a Tungsten inert gas (TIG) device, even under optimal conditions, generates a power density

DOI: 10.1201/9781003492191-7

of 0.1 MW/cm^2, significantly lower than the 10 MW/cm^2 achievable with a low-power laser cladding machine (700 W) [3].

Study [4] provides insights into the maximum power and power density of commonly used laser sources for plating. Notably, CO_2 and Nd:YAG laser sources, with power ranging from 5 to 45 kW, meet the power density requirements for laser coating. Fiber sources, favored for their higher efficiency and power density at lower maximum power, are frequently employed in contemporary applications [5]. This preference aligns with the ever-growing demand for energy-efficient and precise coating processes. It's crucial to acknowledge that certain references propose a power density requirement of 0.7–1 MW/cm^2 for optimal laser cladding [4].

Further exploration into the intricacies of power density, laser sources, and their implications in coating processes can be found in reference [5]. This comprehensive understanding of power density is pivotal for optimizing laser cladding parameters and ensuring its effectiveness in diverse applications.

The localized nature and high power density of laser cladding collectively contribute to its appeal in various industrial domains. The advantages highlighted here are the outcome of extensive research efforts, laying the groundwork for ongoing advancements in laser cladding technology. As laser cladding continues to evolve, these insights provide a foundation for researchers and practitioners to navigate the intricacies of this technology and unlock its full potential in manufacturing and coating applications.

Drawing from existing literature, the inherent locality of the laser cladding process unfolds numerous significant features. This section intensifies the exploration by delving deeper into the advantages linked to this characteristic. Detailed insights and illustrative examples are presented, shedding light on how the localized nature of laser cladding contributes to its efficacy in various industrial applications. Understanding these nuances is pivotal for a comprehensive grasp of the process and its potential implications in diverse manufacturing scenarios.

7.1.1.1 Small HAZ

In the thermal intricacies of laser cladding, specific metal regions undergo metallurgical transformations without reaching the point of melting, forming what is known as the "heat-affected zone" [6]. This region undergoes crucial modifications, including changes in grain size and the formation of fine grains, all attributed to temperature fluctuations during the laser cladding process. Significantly, minimizing the extent of the HAZ is pivotal for mitigating issues such as distortion and residual stress [7]. Understanding the nuances of the HAZ and its impact on metallurgical properties is imperative for optimizing laser cladding parameters and achieving desired outcomes in industrial applications. This insight into the HAZ adds a layer of complexity to the comprehensive understanding of laser cladding, contributing to its efficacy in diverse manufacturing scenarios.

In a comprehensive comparative study, the width of the HAZ took center stage, examining two distinct cladding processes: laser cladding and TIG cladding. The investigation focused on utilizing Inconel 738 as the workpiece material and Inconel 625 as the additive material. Notably, the additive material in laser cladding was in powder form, while in the TIG process, it took the form of wire. Laser cladding operated within a power range of 150–300 W and at a speed of 3–6 mm/min, while the TIG cladding process involved a current range of 40–150 A and a voltage range of 8.3–10.4 V [3].

The study's findings revealed a remarkable distinction in the width of the heat-affected area between the two processes. For the TIG method, the measured width was 200 μm, whereas for the laser cladding method, it was notably narrower at 40 μm. In essence, the HAZ thickness produced by the laser process is approximately one-fifth that of the TIG process [8]. This characteristic of laser cladding, reducing the extent of the HAZ, is of significant importance as it directly translates to diminished distortion and residual stress in the workpiece [3].

This meticulous examination underscores the precision and localized impact of laser cladding, emphasizing its superiority in terms of minimizing the undesirable effects associated with HAZs. The controlled energy deposition and focused nature of the laser beam play a pivotal role in achieving such favorable outcomes. These insights contribute to the growing body of knowledge surrounding laser cladding, providing crucial information for practitioners and researchers in optimizing processes for diverse industrial applications.

7.1.1.2 Low Distortion

The alteration in the geometric shape of a workpiece due to heat application in thermal processes is commonly referred to as "Distortion" [9]. Distortion can be broadly classified into two categories: in-plane distortion and out-of-plane distortion, each encompassing specific types of deformation. In-plane distortion includes transverse contraction, longitudinal contraction, and rotational distortion. Transverse contraction refers to the reduction in width of the workpiece, while longitudinal contraction represents the decrease in length. Rotational distortion refers to the angular misalignment or twisting of the workpiece. On the other hand, out-of-plane distortion comprises angular distortion, longitudinal bending (or bending distortion), and buckling distortion. Angular distortion refers to the tilting or angular displacement of the workpiece. Longitudinal bending, also known as bending distortion, involves the curvature or bending of the workpiece along its length. Buckling distortion refers to the instability leading to the local or global collapse of the workpiece structure [10].

It is worth emphasizing that out-of-plane distortion is frequently considered more significant than other types of distortion, even though all forms of distortion can occur in cladding processes. In line with this, researchers conducted a study to compare the out-of-plane distortion arising from laser and TIG coating methods. The study utilized carbon steel as the workpiece material and 420 stainless steel as the coating powder. The research findings clearly indicated that the laser cladding process resulted in significantly less distortion compared to the TIG method [11].

When the cladding is created with the same hardness at a distance of 50 μm from the surface of the workpiece, a significant distortion is obtained in the sample obtained by the TIG compared to the laser method [11]. It should be noted that studies indicate that distortion has a direct relationship with heat input to the workpiece [12]. In other words, the lower the input heat on the surface, the less distortion is created in the piece. It is worth mentioning that input heat is defined as thermal energy input per unit length [13]. On the other hand, reports indicate that the input heat of laser cladding is much lower than the TIG method. In a study under conditions of the same cladding height, the input heat of the laser method was reported to be one quarter of the TIG method (300 and 1,208 J/m, respectively) [14]. Therefore, because the input heat of the laser cladding process is less, less distortion will be created in the workpiece.

7.1.1.3 High Cooling Rate

The rate at which the temperature decreases following the cladding process is commonly referred to as the "Cooling Rate" [15]. Traditional cladding methods such as TIG, MIG, and others often exhibit a low cooling rate [3]. This can be attributed to the suboptimal distribution and application of energy to the workpiece. In contrast, the laser cladding method offers an advantage in terms of optimal and targeted energy application to a specific area of the workpiece [16]. Optimum energy application entails directing the energy to a specific region to form a molten pool for the cladding process. Conversely, in conventional cladding processes, heat is applied to the entire workpiece, with only a small portion of the energy being utilized for creating the molten pool and carrying out the cladding process [17].

To enhance the depth of comparison, the study [18] offers a comprehensive compilation of cooling rates, specifically focusing on chromium-based alloys in both laser processes and conventional cladding methods. This tabulated data facilitates a detailed examination of the nuanced distinctions in cooling rates, providing valuable insights into how laser processes diverge from traditional cladding techniques. The table serves as a practical tool for researchers and practitioners seeking a thorough understanding of the thermal dynamics involved in laser cladding and its implications for materials, particularly in the context of chromium-based alloys.

Study [18] reveals compelling insights into the cooling rates of various cladding methods for chromium-based alloys. Notably, the laser cladding process stands out with a cooling speed approximately 338 times faster than the oxy-acetylene method and 67 times faster than the TIG method. This substantial discrepancy underscores the pivotal role of high cooling rates in laser cladding, contributing to the formation of a finer microstructure and ultimately enhancing the overall strength of the material [18]. These findings emphasize the superiority of laser cladding in terms of rapid cooling, further solidifying its position as a cutting-edge technology in the realm of material processing.

7.1.2 Challenges in Laser Cladding

As with any advancing technology, laser cladding comes with its own set of disadvantages and distinctive challenges. Despite substantial research efforts aimed at addressing these issues, certain obstacles persist in the process. In the subsequent section, a comprehensive introduction to each of these challenges will be outlined, offering insights into the nuanced complexities that researchers and practitioners face in the field of laser cladding. Understanding and overcoming these challenges is crucial for the continued refinement and successful implementation of laser cladding technology in diverse industrial applications.

7.1.2.1 Complicated Process Controlling

The intricate control of the laser cladding process emerges as a significant challenge, requiring careful consideration [19]. This complexity arises from the dynamic interplay among four phases: solid, liquid, gas, and plasma. Achieving a delicate balance within the confined space of the molten pool adds an additional layer of intricacy, making precise control a formidable task [20]. This challenge necessitates advancements in process monitoring and control mechanisms to optimize outcomes. Addressing these intricacies not only enhances the efficiency of the laser cladding process but also broadens its applicability across diverse industrial sectors. Continuous research efforts, as reflected in references [19,20], play a crucial role in unraveling and overcoming these challenges.

Furthermore, the constrained environment of the molten pool introduces several dynamic forces, including surface tension, the Marangoni effect, and three modes of heat transfer: conduction, convection, and radiation [21]. Each of these factors plays a distinct role in shaping the resulting cladding, contributing significantly to the intricate nature of process control [22]. Surface tension influences the flow of molten material, the Marangoni effect impacts temperature gradients, and the three heat transfer mechanisms govern thermal interactions within the pool [21]. This complex interplay requires sophisticated control strategies to ensure optimal outcomes in laser cladding applications.

Acknowledging the multifaceted challenges, researchers emphasize the imperative for continued investigation and in-depth study to unravel the intricacies and advance the control of the laser cladding process [23]. The challenge lies not only in understanding the individual impacts of these forces but also in orchestrating their harmonious interplay. Addressing these complexities will refine control mechanisms and broaden the applicability of laser cladding across diverse industrial sectors.

7.1.2.2 Environmental Issues

According to the World Health Organization and health organizations in many countries, the environmental effects and associated health concerns pose a significant challenge in laser cladding processes. The laser beam generated during the coating process can have adverse effects on various parts of the body, including the eyes and respiratory system. Depending on the wavelength of the laser beam, it can potentially damage different parts of the eye, such as the cornea, lens, and retina. Although the use of specialized protective eyewear can mitigate these harmful effects to some extent, complete elimination of the risks is not possible. Moreover, the release of powder into the surrounding environment during the powder laser cladding process can have detrimental effects on the respiratory system and may have long-term carcinogenic properties [24]. These environmental and health considerations require careful attention and the implementation of appropriate safety measures to minimize potential risks.

7.2 INDUSTRIAL LASER CLADDING CASES

As mentioned earlier, the laser cladding process finds widespread application in various industries despite the challenges it presents. Its increasing growth in recent years indicates a promising future for this field [25]. The laser cladding process offers advantages that can address different industry challenges efficiently and with high quality. In a general sense, laser cladding applications can be categorized into three main purposes: coating, repair/reconstruction, and rapid prototyping. These applications span across diverse sectors, ranging from automotive [26] to railway lines [27].

One notable advantage of laser coating in this domain is its ability to minimize distortion. It is important to note that laser cladding is not limited by dimensional constraints, enabling the repair of delicate components with complex geometries, such as turbine blades. The attribute "localize" plays a significant role in enabling precise repair or reconstruction [28]. The high working speed characteristic of laser coating makes it particularly suitable for rapid prototyping. The turbine chamber prototyping operation exemplifies the application of laser coating in this area. Notably, the manufacturing time of thin components like turbine housings has been significantly reduced with the utilization of laser coating methods [29].

Furthermore, the laser cladding process offers versatility in terms of the types of materials that can be utilized, including cast iron [30], aluminum [31], steel [32], nickel-based alloys [33], cobalt [34], and more. Based on the types of industries, laser cladding applications can be considered as:

- **Automotive industry**: In the automotive industry, laser cladding finds extensive applications, particularly in coating critical engine components. Engine parts such as cylinder liners, piston rings, valve seats, and turbocharger components benefit from laser cladding processes. This technology contributes to the enhancement of wear resistance, reduction of friction, and overall improvement in the performance and durability of these components. The precision and efficiency of laser cladding make it a valuable technique for advancing the functionality and longevity of automotive parts, aligning with the industry's pursuit of innovative solutions for enhanced performance and reliability.
- **Aerospace industry**: Within the aerospace sector, laser cladding plays a crucial role in the repair and enhancement of critical components such as turbine blades, compressor blades, engine casings, and aerospace alloys. This advanced technology is instrumental in the restoration of worn or damaged parts, ensuring that they regain their original performance and dimensional integrity. Laser cladding's precision and efficiency make it particularly valuable in addressing the demanding requirements of

aerospace applications, where components often face extreme conditions. By seamlessly repairing and improving these vital elements, laser cladding contributes significantly to the reliability, efficiency, and safety of aerospace systems, aligning with the industry's commitment to excellence and cutting-edge solutions.

- **Power generation**: In the realm of power generation systems, laser cladding proves indispensable, finding applications across gas turbines, steam turbines, and hydro turbines. Its utility extends to the repair and enhancement of critical components such as turbine components, boiler tubes, generator shafts, and valve seats. Through precision and efficiency, laser cladding contributes significantly to the restoration of these vital elements, simultaneously improving efficiency and extending the lifespan of power generation components. By addressing wear and damage in a targeted manner, laser cladding plays a pivotal role in ensuring the reliable and sustained operation of power generation systems, aligning with the industry's objectives of enhancing performance, efficiency, and longevity in the crucial domain of energy production.
- **Oil and gas industry**: In the oil and gas sector, laser cladding serves as a vital technology for coating various components, including drilling tools, pump components, valves, and pipes. Its application in this industry is marked by its ability to significantly enhance resistance to wear, corrosion, and erosion in challenging environments. By reinforcing critical components with laser-clad coatings, the technology contributes to increased longevity and reliability of equipment operating in harsh conditions. This approach aligns with the sector's pursuit of durable and high-performance solutions, ensuring the integrity of essential components and promoting the sustained efficiency of operations in oil and gas exploration, drilling, and production activities.
- **Railways**: In the railway industry, laser cladding serves as a valuable technology aimed at improving the performance and longevity of essential components. Components such as axles, wheels, rails, and braking systems benefit from laser cladding applications. This technology plays a crucial role in enhancing wear resistance and reducing maintenance requirements for these critical railway components. By reinforcing these parts with laser-clad coatings, the industry can ensure not only increased durability but also safer and more efficient railway operations. Laser cladding's precision and efficiency contribute to the sector's commitment to reliability and safety, addressing the demanding conditions faced by railway systems and facilitating their optimal functionality.

Indeed, the diverse applications of laser coating underscore its significance across a broad spectrum of industries. From automotive and aerospace to power generation, oil and gas, and railway sectors, laser coating techniques contribute to enhanced performance, prolonged component lifespans, and heightened reliability in systems and equipment. The precision and versatility of laser coating empower industries to address specific challenges, ranging from wear and corrosion resistance to the efficient repair and enhancement of critical components. As a result, laser coating has become an integral technology, fostering innovation and excellence across various domains, ultimately ensuring the sustained efficiency and reliability of industrial processes and infrastructure.

Laser cladding serves a multifaceted role across industries, with applications categorized based on their intended purpose. In coating applications, automotive engine components, aerospace turbine blades, power generation systems, oil and gas tools, and railway elements undergo laser cladding to enhance wear resistance and overall performance. For repair purposes, laser cladding is employed in the automotive sector to restore damaged engine components, while the aerospace, power generation, oil and gas, and railway industries utilize it to extend the lifespan of critical components. In the realm of prototyping, laser cladding plays a pivotal role in the rapid and precise development of automotive, aerospace, power generation, oil and gas, and railway prototypes, allowing for thorough testing and optimization. Across these applications, laser cladding proves indispensable, contributing to enhanced performance, extended component lifespans, and improved reliability in various systems and equipment.

7.3 APPLICATIONS OF LASER BEAM CLADDING IN COATING

Due to the specific working conditions of systems and components, it is often necessary for the surface characteristics to differ from the interior characteristics. For instance, while the surface must withstand wear, erosion, oxidation, and fatigue, the inner part should be able to withstand internal forces. An effective and reliable approach is to create a coating with the desired surface characteristics on the workpiece [35]. In other words, the material for the part is selected based on the required internal characteristics, and then a suitable coating material is chosen and applied to enhance the surface quality. This process is known as "Coating" [36]. To illustrate the application of laser cladding, let's consider the brake disc. Brake discs play a crucial role in reducing the speed of a car by converting kinetic energy into thermal energy through friction. Therefore, characteristics such as wear resistance and high heat transfer are essential considerations when selecting materials for brake discs [37]. Gray cast iron is commonly considered the primary material for brake disc production due to its thermal properties, including a high melting point and a high heat transfer coefficient (1,200°C and 50–72 W/m·K, respectively). These properties accelerate the cooling process of the disc [38]. However, cast iron alloys are known to have low wear resistance [39]. To enhance the life of brake discs, coating technology is widely employed to create a wear- and corrosion-resistant coating.

Traditional methods considered for coating brake discs include chrome plating, high-velocity oxygen fuel, and conventional laser methods. However, chrome plating faces environmental concerns and limited usage due to relevant standards implemented since 2017. The efficiency of coating materials in the high-velocity oxygen fuel method is relatively low, approximately 50%, whereas the laser cladding method exhibits higher efficiency, around 90%. Considering this advantage, recent studies have focused on improving the laser cladding process for brake discs [40].

Although the conventional laser cladding method offers advantages, it also presents challenges when considered for brake disc coating. The typical thickness achieved with conventional laser methods is at least 500 μm, which is considered relatively thick for disc coating [41]. Additionally, the maximum speed of the conventional laser method is 0.5 m/min, which is considered slow for mass-scale production [42]. To address these challenges, a new method called "high-speed laser cladding" was introduced in 2012 to coat brake discs [43]. This method allows for the creation of coatings with a thickness ranging from 50 to 350 μm on the disc surface at a maximum speed of 250 m/min (250 times faster than the conventional laser method). The laser cladding process is currently employed in various studies, and in one study, it was performed using a diode laser source with a power greater than 15 kW and a powder feeding speed of 100 g/min [44].

7.4 APPLICATIONS OF LASER BEAM CLADDING IN REPAIR

Destructive phenomena such as wear, corrosion, oxidation, and fatigue frequently lead to challenges in production, necessitating the replacement of costly components. These issues pose significant hurdles across diverse industries. Laser cladding emerges as a solution with the capacity to repair and restore sensitive and expensive parts, mitigating production challenges. By employing laser cladding, industries can extend the lifespan of critical components, minimizing the need for costly replacements and enhancing overall operational efficiency. In the subsequent section, two practical examples of laser coating repair and restoration will be scrutinized, demonstrating the versatility and effectiveness of laser cladding in addressing issues related to component degradation and ensuring the sustainable functionality of essential industrial parts.

7.4.1 Gas Engine Fuel Nozzle

To optimize energy output from gas turbines, various components within these systems operate under extreme conditions, facing elevated temperatures, pressures, and occasional severe vibrations. In this challenging environment, specific components, such as the turbine fuel nozzle, are susceptible to substantial damage over their operational lifespan. The continuous exposure to high temperatures and pressures, combined with the mechanical stresses induced by severe vibrations, contributes to wear, erosion, and other forms of degradation in critical turbine parts. The degradation of components like the fuel nozzle can adversely impact the overall efficiency and performance of the gas turbine system, emphasizing the importance of advanced maintenance and repair methods, such as laser cladding, to address these issues and extend the operational life of gas turbine components.

The fuel nozzle is responsible for delivering fuel and air into the combustion chamber. The temperature within the combustion chamber of gas turbines can reach up to 1,000°C [45]. Consequently, due to the extreme operating conditions, the air cap of the fuel nozzle can become burned and damaged. Studies indicate that the material used for the fuel nozzle cap is L-605 alloy, also known as Hynes 25. L-605 is a cobalt-based alloy renowned for its excellent oxidation and corrosion resistance. To repair this damaged part, the affected area is initially shaved, followed by the application of cladding using laser technology.

To maximize energy output from gas turbines, many components within these systems are subjected to harsh conditions, including high temperatures, pressures, and severe vibrations. Among these components, the turbine fuel nozzle is particularly susceptible to significant damage throughout its operational lifespan due to the demanding operating environment. Advanced maintenance and repair methods are crucial to address the wear and erosion experienced by components like the fuel nozzle. A study utilizing a 500-W Nd:YAG laser source demonstrated a deposition rate of 5–10 g/min using L-605 powder with particle dimensions ranging from 15 to 45 μm [46]. This laser cladding process proves instrumental in restoring and repairing the fuel nozzle, ensuring its proper functioning, and ultimately extending its operational life. Such studies showcase the efficacy of laser cladding in mitigating issues related to gas turbine component degradation.

To investigate, three different samples were compared: an annealed fuel nozzle, a repaired sample using the laser process, and a sample treated thermally. The annealed sample underwent heat treatment at 732°C for 100 hours, while the heat-treated sample was subjected to a temperature of 1,232°C for 1 hour. The samples were then subjected to a wear test using an abrasion test setup. The test involved a 0.25-inch satellite 20 ball, a vertical force of 150 g, scanning parameters of 66 mm/s scanning speed, and a scanning distance of 3,000 m. The results of the wear test for the three samples are presented in the study [47].

From the study [47], it is evident that the volume of abrasion observed in the laser-repaired sample is significantly lower compared to the annealed sample and is nearly comparable to the heat-treated sample. In other words, the laser method for repairing the part provides wear resistance like that achieved through the heat treatment method without the need for a vacuum furnace and other associated challenges [46]. This indicates that the laser repair process offers a viable alternative to heat treatment for achieving enhanced wear resistance in the fuel nozzle, simplifying the repair procedure while maintaining performance.

7.4.2 Laser Reconstruction of Locomotive Engine Bloc

Internal combustion engines, such as locomotive engines, often experience damage in various parts due to the specific working conditions they endure. The engine block, being one of the critical components, is subject to high temperatures, pressures, and contact between the engine piston and the inner body of the cylinders. Study [48] demonstrates severe burns and damage to the cylinders of such an engine. Notably, the diesel engine in this locomotive weighed 700 kg, was constructed from cast iron, and had a reported power output of 2,500 horsepower. The replacement cost for this engine amounted to at least 14,000 euros, and the process of rebuilding took a considerable amount of time. Consequently, due to the high cost and lengthy replacement time, laser coating is considered a suitable and cost-effective method for repairing such components [49].

In a particular study, the laser coating process was employed to restore the damaged areas within the inner cylinders. The laser device used in the study featured a 3 kW Nd:YAG source with an Icold company head. The selected laser head had the capability to coat surfaces with a minimum inner radius of 26 mm and a maximum depth of 500 mm. The diameter of the repaired engine cylinder in this study was 178 mm, and the coating was applied to a depth of 340 mm. The study reported that by utilizing spiral cutters and maintaining a 90° angle between the surface and the laser head, stainless steel powder with a coating thickness of 1 mm was applied [49].

7.5 APPLICATIONS OF LASER BEAM CLADDING IN PROTOTYPING

Among the emerging applications of laser coating, prototyping complex metal parts stands out. Traditional methods such as machining and molding are often costly, time-consuming, and have low material efficiency. However, laser prototyping offers a faster and more efficient alternative. Reports from the National Center for Manufacturing and Production indicate that prototyping with laser technology can reduce part production time by 40% [50]. This has led to an increasing interest among industrialists who prioritize rapid manufacturing. The process begins with the design of a three-dimensional computer model. The layers of the part are then built up step by step until the final piece is formed [51].

To further exemplify the capabilities of laser prototyping, let's consider the combustion chamber of a helicopter engine produced using laser technology. The aerospace industry has been an early adopter and strong supporter of laser prototyping technology [52]. In the past, plasma welding was used in this context, but it faced challenges such as long production times and high error rates. Laser technology enables the production of hollow parts with optimal geometry in a shorter time frame while maintaining high quality and material efficiency [53]. In a comparison, the combustion chamber of a helicopter engine produced using conventional machining methods took over two months, whereas the same part was produced using laser prototyping in just 7.5 days [54]. Consequently, advanced aviation industries such as Airbus and Boeing have shown significant interest in laser prototyping processes in recent years.

REFERENCES

[1] S. F. Nabavi, M. H. Farshidianfar, A. Farshidianfar, and B. Beidokhti, "Physical-based methodology for prediction of weld bead characteristics in the laser edge welding process," *Optik,* vol. 241, p. 166917, 2021.
[2] W. M. Steen and J. Mazumder, *Laser Material Processing*. Springer, 2010.

[3] M. N. Fesharaki, R. Shoja-Razavi, H. A. Mansouri, and H. Jamali, "Microstructure investigation of Inconel 625 coating obtained by laser cladding and TIG cladding methods," *Surface and Coatings Technology,* vol. 353, pp. 25–31, 2018.
[4] E. Toyserkani, A. Khajepour, and S. F. Corbin, *Laser Cladding.* CRC Press, 2004.
[5] G. Turichin, E. Zemlyakov, E. Y. Pozdeeva, J. Tuominen, and P. Vuoristo, "Technological possibilities of laser cladding with the help of powerful fiber lasers," *Metal Science and Heat Treatment,* vol. 54, pp. 139–144, 2012.
[6] F. Khodabakhshi, M. Farshidianfar, A. Gerlich, M. Nosko, V. Trembošová, and A. Khajepour, "Effects of laser additive manufacturing on microstructure and crystallographic texture of austenitic and martensitic stainless steels," *Additive Manufacturing,* vol. 31, p. 100915, 2020.
[7] M. H. Farshidianfar, A. Khajepour, and A. Gerlich, "Real-time control of microstructure in laser additive manufacturing," *The International Journal of Advanced Manufacturing Technology,* vol. 82, pp. 1173–1186, 2016.
[8] M. H. Farshidianfar, *Control of Microstructure in Laser Additive Manufacturing.* University of Waterloo, 2014.
[9] W. D. Callister Jr and D. G. Rethwisch, *Fundamentals of Materials Science and Engineering: An Integrated Approach.* John Wiley & Sons, 2020.
[10] R. B. Hetnarski, *Encyclopedia of Thermal Stresses.* Springer, 2014.
[11] M. Soodi, "Laser cladding compared with TIG welding to repair and refurbish railway axles," In: *Conference on Railway Engineering,* Wellington, NZ. Engineers Australia Wellington, NZ, pp. 516–523, 2010.
[12] F. Khodabakhshi, M. Farshidianfar, A. Gerlich, M. Nosko, V. Trembošová, and A. Khajepour, "Microstructure, strain-rate sensitivity, work hardening, and fracture behavior of laser additive manufactured austenitic and martensitic stainless steel structures," *Materials Science and Engineering: A,* vol. 756, pp. 545–561, 2019.
[13] Y.-C. Kim, M. Hirohata, M. Murakami, and K. Inose, "Effects of heat input ratio of laser-arc hybrid welding on welding distortion and residual stress," *Welding International,* vol. 29, no. 4, pp. 245–253, 2015.
[14] W. M. Steen and W. M. Steen, *Laser Surface Treatment.* Springer, 2003.
[15] M. H. Farshidianfar, A. Khajepour, and A. P. Gerlich, "Effect of real-time cooling rate on microstructure in laser additive manufacturing," *Journal of Materials Processing Technology,* vol. 231, pp. 468–478, 2016.
[16] Y. Huang et al., "Rapid prediction of real-time thermal characteristics, solidification parameters and microstructure in laser directed energy deposition (powder-fed additive manufacturing)," *Journal of Materials Processing Technology,* vol. 274, p. 116286, 2019.
[17] G. Muvvala, D. P. Karmakar, and A. K. Nath, "Online monitoring of thermo-cycles and its correlation with microstructure in laser cladding of nickel based super alloy," *Optics and Lasers in Engineering,* vol. 88, pp. 139–152, 2017.
[18] J. Powell, P. Henry, and W. Steen, "Laser cladding with preplaced powder: analysis of thermal cycling and dilution effects," *Surface Engineering,* vol. 4, no. 2, pp. 141–149, 1988.
[19] M. H. Farshidianfar, A. Khajepour, S. Khosravani, and A. Gelrich, "Clad height control in laser cladding using a nonlinear optimal output tracking controller," *International Congress on Applications of Lasers & Electro-Optics*, vol. 2013, no. 1, pp. 470–479, 2013.
[20] M. H. Farshidianfar, A. Khajepouhor, and A. Gerlich, "Real-time monitoring and prediction of martensite formation and hardening depth during laser heat treatment," *Surface and Coatings Technology,* vol. 315, pp. 326–334, 2017.

[21] C. Panwisawas et al., "Keyhole formation and thermal fluid flow-induced porosity during laser fusion welding in titanium alloys: Experimental and modelling," *Acta Materialia,* vol. 126, pp. 251–263, 2017.
[22] M. H. Farshidianfar, F. Khodabakhshi, A. Khajepour, and A. P. Gerlich, "Closed-loop control of microstructure and mechanical properties in additive manufacturing by directed energy deposition," *Materials Science and Engineering: A,* vol. 803, p. 140483, 2021.
[23] M. H. Farshidianfar, "Real-time closed-loop control of microstructure and geometry in laser materials processing," PhD Thesis, University of Waterloo, 2017.
[24] W. Cai, J. Wang, P. Jiang, L. Cao, G. Mi, and Q. Zhou, "Application of sensing techniques and artificial intelligence-based methods to laser welding real-time monitoring: A critical review of recent literature," *Journal of Manufacturing Systems,* vol. 57, pp. 1–18, 2020.
[25] L. Zhu et al., "Recent research and development status of laser cladding: A review," *Optics & Laser Technology,* vol. 138, p. 106915, 2021.
[26] A. Singh, A. Ramakrishnan, and G. P. Dinda, "Direct laser metal deposition of eutectic Al-Si alloy for automotive applications," In: *TMS 2017 146th Annual Meeting & Exhibition Supplemental Proceedings*, Springer, pp. 71–80, 2017.
[27] L. Meng, B. Zhu, C. Xian, X. Zeng, Q. Hu, and D. Wang, "Comparison on the wear properties and rolling contact fatigue damage behaviors of rails by laser cladding and laser-induction hybrid cladding," *Wear,* vol. 458, p. 203421, 2020.
[28] J. R. Lawrence, *Advances in Laser Materials Processing: Technology, Research and Applications*. Woodhead Publishing, 2017.
[29] M. Meboldt and C. Klahn, *Industrializing Additive Manufacturing-Proceedings of Additive Manufacturing in Products and Applications-AMPA2017.* Springer, 2017.
[30] B. Graf, A. Gumenyuk, and M. Rethmeier, "Laser metal deposition as repair technology for stainless steel and titanium alloys," *Physics Procedia,* vol. 39, pp. 376–381, 2012.
[31] M. Froend, V. Ventzke, S. Riekehr, N. Kashaev, B. Klusemann, and J. Enz, "Microstructure and microhardness of wire-based laser metal deposited AA5087 using an Ytterbium fibre laser," *Materials Characterization,* vol. 143, pp. 59–67, 2018.
[32] M. H. Farshidianfar, F. Khodabakhshi, A. Khajepour, and A. Gerlich, "Closed-loop deposition of martensitic stainless steel during laser additive manufacturing to control microstructure and mechanical properties," *Optics and Lasers in Engineering,* vol. 145, p. 106680, 2021.
[33] J. Liu, H. Liu, X. Tian, H. Yang, and J. Hao, "Microstructural evolution and corrosion properties of Ni-based alloy coatings fabricated by multi-layer laser cladding on cast iron," *Journal of Alloys and Compounds,* vol. 822, p. 153708, 2020.
[34] Y. Zou, B. Ma, H. Cui, F. Lu, and P. Xu, "Microstructure, wear, and oxidation resistance of nanostructured carbide-strengthened cobalt-based composite coatings on Invar alloys by laser cladding," *Surface and Coatings Technology,* vol. 381, p. 125188, 2020.
[35] R. Vilar, "Laser cladding," *Journal of laser applications,* vol. 11, no. 2, pp. 64–79, 1999.
[36] A. Matthews and D. S. Rickerby, *Advanced Surface Coatings: A Handbook of Surface Engineering.* Springer, 1991.
[37] A. Rashid, "Overview of disc brakes and related phenomena-a review," *International Journal of Vehicle Noise and Vibration,* vol. 10, no. 4, pp. 257–301, 2014.
[38] O. Aranke, W. Algenaid, S. Awe, and S. Joshi, "Coatings for automotive gray cast iron brake discs: A review," *Coatings,* vol. 9, no. 9, p. 552, 2019.

[39] M. Shin, K. Cho, S. Kim, and H. Jang, "Friction instability induced by corrosion of gray iron brake discs," *Tribology Letters,* vol. 37, pp. 149–157, 2010.
[40] T. Schopphoven, J. H. Schleifenbaum, S. Tharmakulasingam, and O. Schulte, "Setting Sights on a 3D Process: Extreme high-speed laser material deposition, or EHLA, is now being developed into an additive manufacturing process," *Photonics Views,* vol. 16, no. 5, pp. 64–68, 2019.
[41] M. H. Farshidianfar, A. Khajepour, M. Zeinali, and A. Gelrich, "System identification and height control of laser cladding using adaptive neuro-fuzzy inference systems," *International Congress on Applications of Lasers & Electro-Optics,* vol. 2013, no. 1, pp. 615–623, 2013.
[42] F. ILT, "Coating through High-speed Laser Material Deposition," 2014.
[43] T. Schopphoven, A. Gasser, and G. Backes, "EHLA: Extreme High-Speed Laser Material Deposition: Economical and effective protection against corrosion and wear," *Laser Technik Journal,* vol. 14, no. 4, pp. 26–29, 2017.
[44] M. Barbosa, R. Bischoff, W. Strauß, H. Hillig, S. Nowotny, and C. Leyens, "Less CO2 and Fine Dust Emissions in Automotive: High-power laser cladding as a cost-effective rotor coating solution," *PhotonicsViews,* vol. 17, no. 4, pp. 46–49, 2020.
[45] B. W. Lagow, "Materials selection in gas turbine engine design and the role of low thermal expansion materials," *Jom,* vol. 68, no. 11, pp. 2770–2775, 2016.
[46] M. H. Farshidianfar, S. F. Eshraghi, S. F. Nabavi, and A. Farshidianfar, "A review on the presence of lasers in the processes of coating, repair, and prototyping," *Metallurgical Engineering,* vol. 24, no. 4, pp. 325–342, 2022.
[47] L. Xue, M. Donovan, Y. Li, J. Chen, S. Wang, and G. Campbell, "Integrated rapid 3D mapping and laser additive repair of gas turbine engine components," In: *International Congress on Applications of Lasers & Electro-Optics.* AIP Publishing, pp. 318–325, 2013.
[48] S. Kalawrytinos and H. Desmecht, *Rejuvenating Engines with Rotating Internal Processing Heads: Efficient Repair of Worn Cylinder Bores Through Laser Cladding.* Wiley Online Library, vol. 7, pp. 32–35, 2010.
[49] A. S. Kalawrytinos and H. Desmecht, "Laser cladding of worn cylinder bores," 2011. https://www.laserfocusworld.com/industrial-laser-solutions/article/14216545/laser-cladding-of-worn-cylinder-bores
[50] J. Mazumder, D. Dutta, N. Kikuchi, and A. Ghosh, "Closed loop direct metal deposition: art to part," *Optics and Lasers in Engineering,* vol. 34, no. 4-6, pp. 397–414, 2000.
[51] M. Touri, F. Kabirian, M. Saadati, S. Ramakrishna, and M. Mozafari, "Additive manufacturing of biomaterials– the evolution of rapid prototyping," *Advanced Engineering Materials,* vol. 21, no. 2, p. 1800511, 2019.
[52] I. G. Ian Gibson, *Additive Manufacturing Technologies 3D Printing, Rapid Prototyping, and Direct Digital Manufacturing.* Springer, 2015.
[53] R. Liu, Z. Wang, T. Sparks, F. Liou, and J. Newkirk, "Aerospace applications of laser additive manufacturing," In: *Laser Additive Manufacturing.* Elsevier, pp. 351–371, 2017.
[54] C. Hauser, "Case Study: Laser powder metal deposition manufacturing of complex real Parts," TWI https://docplayer. net/48458850-Case-study-laser-powder-metal-depositionmanufacturing-of-complex-real-parts. html, 2014.

8 Additive Manufacturing

8.1 INTRODUCTION

The advent of digital technology and its displacement of analog processes have ushered in significant advancements in fields such as communication, imaging, architecture, and engineering. In a similar vein, additive manufacturing represents a transformative force in the realm of manufacturing, much like analog processes did for the digital world. By leveraging additive manufacturing, the production of parts and the overall manufacturing landscape have undergone remarkable transformations. This technology has effectively provided the construction and production sectors with the flexibility and efficiency previously enjoyed by the digital realm.

Additive manufacturing encompasses various industrial processes, including rapid prototyping and 3D printing. Rapid prototyping, or RP, refers to a set of industrial processes where the system or components are produced prior to the final sample. The focus of these methods lies in high-speed production, resulting in a basic model or sample. Consequently, the prototype produced may differ in quality from the final sample. While software engineers attribute a different concept to "Rapid prototyping," in the manufacturing world, it denotes the technology that produces a physical sample directly from digital model data. Although initially employed solely for prototyping purposes, this technology is now utilized in the production of final parts for various applications.

As the applications of RP technology have expanded extensively, users have come to realize that the term "Rapid prototyping" no longer adequately captures the progress made in recent years. With improved output quality correlating to enhanced final product quality, efforts have focused on upgrading these machines. Consequently, researchers and manufacturers have sought to produce the final product directly using this method, rendering the term "Prototyping" inadequate. The term "Additive manufacturing" emerged as the new designation for this process, as recognized by the ASTM (American Society for Testing and Materials) standard forum, although debates persist in certain circles. This chapter delves into the evolution and significance of additive manufacturing, showcasing its immense potential for transforming the manufacturing landscape. By examining the development from RP to additive manufacturing, the chapter highlights the continuous progress made in achieving high-quality, end-use parts directly from digital models.

Additive manufacturing, also known as 3D printing, revolutionizes the production process by utilizing computer-aided design (CAD) software or 3D scanners to translate data and commands. With precise geometry based on this information, the hardware layers materials, injecting them to create specific parts. In contrast,

DOI: 10.1201/9781003492191-8

traditional manufacturing methods often require post-processing techniques like milling, machining, and shaping to remove excess components.

The underlying principle of additive manufacturing, or AM, is based on generating a 3D model using CAD software. This model can be directly translated into a physical product without the need for extensive planning. Although the reality of the production process is more intricate, additive manufacturing offers a simplified approach compared to traditional methods. In traditional manufacturing, achieving the desired dimensional tolerance often necessitates using multiple separate devices, whereas AM only requires basic dimensional specifications, understanding of the AM machine's operation, and knowledge of fabrication materials.

The key principle of additive manufacturing (AM) lies in its layer-by-layer material addition process, where each layer represents a small cross-section and thickness of the original product. The order in which these layers are created is determined by CAD data, ensuring a close approximation to the desired model. Various stages, including the CAD model, faceted model, virtual cut model, tool movement path, successive layers, and the final cup-shaped product, are considered in the AM process. It clearly demonstrates that the thickness of each layer plays a crucial role in achieving accurate results that closely match the software model. The smaller the layer thickness, the higher the accuracy of the manufactured part.

Additive manufacturing (AM), despite appearing relatively new to some, has been in existence for several decades. Its continuous development and impressive capabilities have enabled the production of parts with complex geometries while maintaining a simplified manufacturing process. This has opened a wide range of opportunities for innovation in the field of incremental construction. What sets AM apart is its ability to offer engineers unprecedented advantages in terms of product performance, reliability, and production speed. Unlike traditional manufacturing methods, AM allows for the creation of highly intricate and customized designs that were previously difficult or even impossible to produce. By adding material layer by layer, engineers can achieve complex geometries, intricate internal structures, and optimized designs, all while minimizing waste material.

Moreover, AM facilitates the production of parts with superior performance and reliability. The ability to precisely control the deposition of material enables the creation of parts with tailored properties, such as improved strength-to-weight ratios and enhanced functional integration. This level of customization can result in optimized performance, reduced weight, and increased durability, leading to products that outperform those made using traditional manufacturing techniques. Another notable advantage of AM is its high production speed. By directly translating digital designs into physical objects, AM eliminates the need for time-consuming tooling and setup processes. This streamlined workflow allows for RP, on-demand manufacturing, and reduced time-to-market for new products. Additionally, the ability to produce complex assemblies as a single integrated part can significantly reduce the assembly time and labor required, further enhancing production efficiency.

In summary, AM offers significant advantages over traditional manufacturing methods. Its ability to create complex geometries, enhance product performance, ensure high reliability, and accelerate production speed makes it a transformative technology in various industries. By harnessing these unique benefits, engineers and manufacturers can unlock new possibilities for design innovation, customization, and efficiency in their production processes.

8.2 APPLICATIONS OF AM

The advent of AM has had a transformative impact on the design of parts across numerous industries. This revolutionary technology has paved the way for remarkable advancements and applications. To provide a comprehensive overview of its widespread adoption, the study [1] compiles industry-specific statistics from 2017, illustrating the utilization of AM in sectors such as aerospace, automotive, industrial machinery, architecture, military, academia, medicine, and more. The data reveals that the largest shares of AM implementation can be found in the fields of industrial machinery, aerospace, automotive, consumer/electronics, and medical products, respectively.

To further explore the breadth of AM's applications, studies [2–4] showcase diverse examples within these industries. These visual representations highlight the versatility of AM by demonstrating its ability to fabricate complex components, produce prototypes, enable customization, and even contribute to the development of medical implants, among other possibilities. By delving into these industry-specific use cases, one can gain a deeper appreciation for the transformative power of AM. This technology unlocks new levels of design freedom, efficiency, and innovation, pushing the boundaries of what is achievable in manufacturing. It is through these practical implementations that the true potential and impact of AM become abundantly clear, revolutionizing industries and opening up new realms of possibility for design and production.

The study [1] provides compelling evidence of the substantial growth of AM technology. The sales data pertaining to AM services, software, materials, and equipment from 2014 to 2023 clearly demonstrates a remarkable upward trend. According to the report, the market size for AM is projected to soar from 400 million dollars in 2014 to an impressive 1,200 million dollars by 2023, indicating a threefold increase. This substantial growth reflects the increasing recognition and adoption of AM across industries.

The significance of AM cannot be overstated, as evidenced by its robust market growth. It has become a technology that cannot be ignored, given its immense potential and impact. Recognizing its importance, the following section will delve into a comprehensive exploration of the advantages offered by the AM process. By understanding these benefits, one can fully grasp the transformative power of AM and its implications for the future of manufacturing and design.

8.3 ADVANTAGES OF AM

AM has been hailed as a revolutionary technology by researchers and industrialists, with some even suggesting that the world would be significantly behind if AM had not been pursued [5]. This non-destructive manufacturing technology has transformed the realm of product design and spurred enthusiasm among industries and researchers alike. This raises the question: What are the distinctive advantages and characteristics of AM that have garnered such fervor?

To address these inquiries, it is helpful to categorize part production based on the underlying technology. Study [6] provides a classification into five broad categories: (1) joining, (2) separation, (3) shaping (casting/molding), (4) subtractive manufacturing (machining), and (5) incremental construction. According to a study [6], joining technology involves processes in which two or more parts are connected to create a new part. Welding and assembly are examples of this method. In contrast, the separation or division process entails creating a new part by separating or disassembling existing parts, such as through the use of a saw.

These categories highlight the diverse approaches to part production and emphasize the uniqueness of AM. Unlike traditional methods that primarily fall into joining, separation, or shaping techniques, AM enables a novel approach known as incremental construction. By adding material layer by layer, it offers unprecedented design freedom and the ability to fabricate complex geometries, internal structures, and customized components with minimal waste.

The advantages of AM extend beyond its ability to create intricate parts. It also enables RP, on-demand manufacturing, reduced time-to-market, and cost-efficient production. The technology's versatility and potential applications have captivated industries across various sectors, including aerospace, automotive, healthcare, and consumer goods. By understanding the advantages and distinct characteristics of AM, one can appreciate why it has garnered such enthusiasm and why it is regarded as a game-changing technology in the world of manufacturing and design.

Formative manufacturing technology involves applying another workpiece to shape the desired part [7]. This method encompasses processes such as forming, heat treatment, and cryogenic cooling. It is particularly suitable for mass production of a specific part, where the manufacturer invests in the necessary tools to produce large quantities of the part at a low cost in the future [7,8]. Subtractive manufacturing involves the gradual removal of materials to achieve the desired part geometry. Over the past two decades, lean manufacturing technologies have undergone significant transformations. Traditional codes like M and G have been replaced by 3D modeling software. The new generation of computer numerical control (CNC) machines is integrated with CAD and computer-aided manufacturing systems [8]. Subtractive manufacturing is typically used for parts with relatively simple geometries and produced in medium volumes [7].

Study [7] illustrates AM, a process where materials are added and injected step by step to create the final part. This category of production methods includes casting, injection molding, and RP [8]. AM is well-suited for producing complex parts that cannot be easily manufactured using shaping or subtractive manufacturing methods. It is also ideal when fast production of prototypes or samples is required [7]. Investigating these different production technologies helps in appreciating the unique advantages and applications of AM. It offers unparalleled design freedom, the ability to produce complex geometries, and RP, setting it apart from traditional shaping and subtractive manufacturing techniques.

The presented production methods can be compared from different points of view. Considering that one of the most important criteria is the comparison based on the complexity and cost of production, in the study [9], two SLM methods (one of the most important branches of the AM method) and CNC (as a method based on subtractive manufacturing) have been compared from these two points of view. It is worth mentioning that the amount of 20–60 represents the least and the most complexity [9]. According to this figure, for low complexity (below 48), the use of the CNC method has a lower production cost than the SLM method, so it is more economical. While increasing the complexity of the work piece, the incremental manufacturing method will be more affordable. Also, for complexities of more than 50, it is practically not possible to produce parts by the CNC method. The study [9] shows the lathe tool produced by Sandvik as the result of two methods of additive and subtractive manufacturing.

The study [9] showcases the ability of AM to fabricate turning tools with complex geometries. The AM process not only enables the production of intricate shapes but also facilitates integrated manufacturing. This integration leads to a reduction in production time for parts. Furthermore, Study [9] presents an example from Siemens, illustrating the industrial gas turbine produced using both additive and subtractive manufacturing methods. According to the company's report, the traditional subtractive manufacturing method requires 13 components, 18 welding lines, and a maximum production time of 26 weeks. In contrast, the AM method allows for the integrated production of the part within a mere 3 weeks. This significant reduction in production time is achieved by utilizing the AM approach, which minimizes the number of components required.

These examples highlight the advantages of AM in terms of complex geometry production, integrated manufacturing, and reduced production time. By consolidating multiple components into a single integrated part, AM not only simplifies the manufacturing process but also accelerates the overall production timeline. This demonstrates the transformative potential of AM in streamlining production and increasing efficiency.

Comparing conventional methods like subtractive manufacturing and forming with AM often involves considering the duration and number of production processes. The study [9] provides a visual representation of the production steps in incremental manufacturing, comparing it with subtractive manufacturing in terms of the number of processes, duration, and efficiency. According to the study [9], the subtractive manufacturing process requires eight processes and takes approximately 9 months to complete in order to produce the desired parts. On the other hand, AM stands out by utilizing only three processes for part production. Moreover, AM allows for continuous production, meaning parts can be produced in real-time as needed. This comparison highlights the efficiency and time-saving benefits of AM over conventional methods. By streamlining the production process and reducing the number of required steps, AM offers significant advantages in terms of speed, flexibility, and responsiveness to production demands.

The visualization in the study [9] underscores the potential for AM to revolutionize manufacturing processes, enabling faster and more efficient production of parts compared to traditional subtractive manufacturing and forming methods.

In a comprehensive study, a comparison was conducted between AM, subtractive manufacturing, and forming methods to produce the Gardan Double H-Hook industrial part. The Double H-Hook is a vital component of a car's power transmission system, responsible for transferring torque from the gearbox to the differential. Specifically, the H-hook joint, which forms part of the double girdle or central hook, plays a crucial role in transmitting force with a constant angular velocity. Traditionally, H-hook joints have been manufactured using cast iron casting and machining processes. However, in the study mentioned, these different manufacturing methods were examined to evaluate their suitability for producing the H-hook joint.

According to the study [10], the H-hook joint used in the study had a volume of 262.88 m^3 and was fabricated using AISI 630 steel, known for its durability and strength [10]. This specific case study aimed to explore the feasibility and potential benefits of AM, subtractive manufacturing, and forming methods for manufacturing the H-hook joint. Further details regarding the findings and conclusions drawn from this comparative analysis would require a more in-depth examination of the study itself.

The study comparing different manufacturing methods for the Gardan Double H-Hook industrial part yielded insightful results. It was observed that the casting process had the longest production time, requiring approximately 55 hours. In contrast, the jet binder (a type of AM) and powder bed melting (another type of AM) methods had significantly shorter production times of approximately 15.2 and 6 hours, respectively. Thus, in terms of production time, the powder bed melting process emerged as the most efficient option.

From an environmental perspective, the machining process was found to have the highest water consumption. Additionally, the energy requirements for machining were significantly higher compared to other methods. Consequently, the study concluded that the machining process (a subtractive manufacturing method) is not desirable due to its high energy consumption and water usage. Similarly, the casting process was noted to be time-consuming and associated with lower production speed.

Based on these findings, the AM process was deemed the most suitable method for producing the Gardan Double H-Hook. AM offered several advantages, including high production speed, lower energy requirements, and greater environmental friendliness. Taking all factors into account, the study determined that AM was the optimal choice for efficiently producing the H-Hook industrial part [10].

To facilitate better understanding, a comprehensive comparison between the features of the CNC subtractive manufacturing process and AM is compiled in the study [3]. This study highlights the notable differences between the two processes from various perspectives. AM, in contrast to CNC subtractive manufacturing, offers advantages such as increased production speed, improved part quality, and lower cost. Furthermore, as mentioned in earlier sections, AM enables the production of complex parts that would be challenging or impossible to achieve using traditional methods [3].

The study [11] presents a comparison of the unit price based on complexity level and production volume for AM and conventional manufacturing methods. The figure illustrates that in scenarios involving medium to low production volume and medium to high complexity, the utilization of AM is recommended, as it offers cost advantages and efficiency benefits over conventional methods [11].

It is important to note that each manufacturing process has its own set of advantages and limitations, and AM is no exception. While AM brings numerous benefits, it also possesses certain limitations. These limitations are summarized in the study [1], providing a comprehensive overview of the challenges and considerations associated with AM. By understanding both the advantages and limitations, stakeholders can make informed decisions and leverage AM effectively in their specific applications and contexts.

8.4 TYPES OF AM

According to the ISO/ASTM 52900:2015 standard, commercialized AM processes are categorized into seven distinct groups. These categories are defined as follows: direct energy deposition (DED), vat photopolymerization (VP), powder bed fusion (PBF), binder jetting (BJ), material jetting, sheet lamination (SL), and material extrusion (ME) [1]. The study [3] provides an overview of the technologies and materials associated with these AM processes. The table highlights the specific technologies employed in each category, while the figure visually represents the corresponding materials used in these processes. This information helps to better understand the variety and diversity of AM technologies and the materials utilized within each category.

By categorizing and defining these AM processes, the ISO/ASTM 52900:2015 standard provides a framework for classifying and communicating the various methodologies and techniques employed in the field. This standardized approach facilitates better understanding, comparability, and collaboration within the AM industry.

From an alternative standpoint, AM research papers often classify AM processes based on the state of the secondary material, following the ASTM F42T standard. These processes can be categorized into three types: solid, liquid, and powder. Studies [3] and [1] provide a visual representation and comparison of this classification. In these studies, solid wires, molten metal, and metal powder are classified as solid, melt, and powder, respectively. Another perspective divides AM processes into categories based on the primary material form, including powder, wire, plate, rod, dispersion, filament, and bullet, as illustrated in the study [9]. This chapter delves further into specific AM processes. In the melt-based category, laser powder bed melting and direct energy injection (DED) processes will be explored. The solid-based category encompasses jet binder, extrusion-based injection, cold spray, and friction stir welding, all of which will be presented in detail.

8.5 POWDER BED FUSION

The PBF technology, which encompasses two methods, laser (laser powder bed melting) and electron beam (EBM), traces its origins back to the invention of Ross Housholder in 1979 [9]. Housholder drew inspiration from observing layered sand on the beach during his daily commute in Las Vegas. He conceived the idea of simultaneously placing sand and cement in square grids. Although he considered using a laser in his invention, the unavailability of laser equipment prevented him from pursuing this aspect further.

The study [12] provides a visual representation of Housholder's invention. Eventually, this invention was sold to the DTM company (formerly Selective Sintering Company or SLS), founded by Carl Deckard and his associates. Notably, SLS is based on Housholder's invention, which received 260 citations from various sources before its expiration [9].

Following this, Deckard embarked on the task of integrating a computer-controlled laser. He utilized a Commodore 64 computer model and designed a board that enabled 4 kb of coding. Once the control parameters were finalized, at the suggestion of Dr. Beaman, Deckard presented a portion of this research in his master's thesis. Subsequently, in 1986, Deckard decided to continue his collaboration with Dr. Beaman as a doctoral student at the University of Texas to complete the project. With a $30,000 grant from the National Science Foundation (NSF), they successfully built a research device named "Betsy". Equipped with electrical panels, rollers, and associated controllers, Betsy enabled the layer-by-layer production of laboratory parts. Study [9] showcases the Betsy power supply system and one of the first parts produced by Betsy. This device served as one of the earliest devices in the field, contributing to a deeper understanding of the physics of the powder bed melting process and its applications. Further details about the device and its application will be presented and reviewed in subsequent sections.

8.5.1 Process physics

The powder bed melting process involves selectively melting layers of a powder bed using a heat source, such as a laser or electron beam, to create a three-dimensional solid object [13]. As depicted in the study [9], the powder bed is provided by the powder tank and displayed on the screen. The laser beam passes through a scanning mirror and lenses, targeting specific areas of the powder bed based on the part's design, causing the powder to melt. Once the first layer solidifies, subsequent layers are added according to the part's plan, following the same process as the initial layer. This layer-by-layer approach continues until the complete piece is produced. It is important to note that for hollow parts, supports are incorporated during the manufacturing process. These supports provide stability and structure during production and can be removed at the end of the process.

8.5.1.1 Selective Laser Sintering (SLS)

In the selective laser sintering (SLS) process, a view of which is considered in the study [4], powder mixed with two elements with a high melting temperature and a low melting temperature is often used. In the SLS process, the powder with a low melting temperature is melted on the surface, and it connects with unmelted powders with a high melting temperature. The powders are guided to the bed by the roller. According to the three-dimensional CAD data, the laser beam is guided selectively onto the powder bed by the scanner. By changing the height of the powder bed according to the required thickness of each layer, the next layer will be formed on the first sintered layer. Again, in this layer, the laser beam is selectively irradiated on the workpiece, and the second layer is connected to the first layer. This process will continue until the complete construction of the part is completed. A CO_2 laser beam is often used in this process [3].

8.5.1.2 Selective Laser Melting

The SLS process and the SLM process exhibit similar performance, leading some to consider them essentially the same. However, there are distinct differences between the two. In the SLS process, powder particles are sintered and undergo surface melting. In contrast, the SLM process involves the complete melting of the powder. The SLM method offers several advantages. It reduces production time for complex parts and results in a finer microstructure within the manufactured components due to rapid cooling. The mechanical properties of parts produced using the SLM method are particularly impressive, making it a preferred choice for sensitive industries such as aerospace, automotive, and medicine [3].

8.5.1.3 Direct Metal Laser Sintering (DMLS)

Based on the study [14], it can be observed that the direct metal laser sintering (DMLS) process shares similarities with the selective laser melting (SLM) process in terms of process physics. The DMLS method is utilized for the production of prototypes and graded materials (FGM). Notably, this method finds applications in the aerospace [3] and medical [15] sectors due to its advantages, including reduced raw material requirements and lower production costs. In the DMLS process, each layer typically has a thickness ranging from 20 to 60 μm, and the power used is typically around 200 W. It is worth mentioning that although DMLS, SLS, and SLM processes share fundamental process physics, they differ in terms of the type of powder employed [14].

8.5.1.4 Selective Heat Sintering (SHS)

In the selective heat sintering process, polymer powder particles are transferred from the feeding tank to the production tank using a roller mechanism. The system tool's movement path is determined based on the CAD data, considering the part specifications and boundary conditions with the nozzle. Heat is then radiated onto the polymer powder using a cost-effective heater. The polymer particles are heated to a temperature below their melting point and sintered at a specific temperature. Through layer-by-layer sintering based on the boundary conditions, a three-dimensional piece is created. This AM method offers several advantages. It allows for the use of high- and low-density polyethylenes, polyamides, polycarbonates, and other suitable materials. Additionally, the process is relatively inexpensive compared to methods involving electron beams or laser beams.

8.5.1.5 Electron Beam Melting (EBM)

The electron beam process is an AM method that was initially introduced in 1997. It belongs to a category of technologies where the melting process is facilitated by an electron beam serving as the energy source. The energy required for the electron beam is generated through the control of the tungsten filament and coil. With the electron gun being computer-controlled, a three-dimensional metal powder piece is produced. Study [3] depicts the schematic representation of this process. Similar to other AM processes, the EBM process involves melting layers of metal powder to create the desired parts [3].

8.5.1.6 Comparison of the PBF Process

In the preceding sections, we have introduced various powder bed melting processes. To provide a summary, this section aims to compare these processes. Studies [3,14] present an overview of these processes, including the year of their first invention, the inventor, the manufacturing company and its headquarters, the initial system built, the type of secondary material, the year of market entry, and the manufacturing company associated with each process. Among the processes listed in studies [3,14], SLM and EBM have garnered significant attention from industrialists in recent years. Studies indicate that the fabrication rate of the EBM process, which benefits from higher energy density and faster scanning speeds, exceeds 80 cm^3/h. Conversely, the surface quality and dimensional tolerance of parts produced by the EBM process may be inferior to those created by the SLM process [16].

Both the SLM and EBM processes generate residual stresses due to the rapid heating and cooling of the powder layers. However, in the EBM process, the high temperature within the manufacturing chamber (ranging from 700°C to 900°C) allows for the possibility of preheating and reducing the temperature gradient of the powder. This, in turn, helps in minimize residual stresses. Moreover, since the entire EBM process takes place within the chamber, there is a reduction in heat loss and enhanced protection of the molten pool, leading to improved process control [16]. In the SLM process, gas protection is achieved by using argon, which is injected into the powder bed. Consequently, it can be expected that parts produced using the EBM process will have a lower oxygen content compared to those manufactured via SLM [17].

In the study [18], the microstructure and mechanical properties of a Ti-6Al-4V alloy part manufactured using electron beam melting (EBM) and selective laser melting (SLM) processes were compared. The study [18] presents SEM images of the powders utilized in the SLM and EBM processes. A noticeable distinction is observed between the powder sizes, with the SLM process employing smaller particles (average 35 μm) compared to the EBM process (average 77 μm).

The EBM process parameters consisted of a voltage of 60 kV, a beam size of 200 μm, a vacuum chamber of 10–3 microbars, and a speed of 1,500 mm/s. On the other hand, the SLM process parameters were as follows: laser beam diameter of 40 μm, speed of 1,000 mm/s, and power of 200 W. The reported input energy density for SLM and EBM was 40 and 76 J/mm^3, respectively. Both the EBM and SLM processes employed a layer thickness of 0.05 mm.

Microstructural analysis of the AM samples is displayed in the study [18], depicting the outcomes of the SLM and EBM processes, respectively. It can be observed that the EBM-produced part exhibits an α lamella structure with some β phase, while the sample generated through the SLM process consists of thin and elongated martensitic chips. Moreover, the study found that the strength of the SLM samples (1,400 MPa) surpassed that of the EBM samples (1,000 MPa). However, the ductility of the SLM samples was reported to be lower than that produced through the EBM method.

Considering the study [18], the pores observed in the SLM process are generally irregular in shape, whereas the porosity resulting from the EBM process appears spherical. Furthermore, the porosity of the EBM process ($4.87 \times 10^{-5 \mu m}$) was lower than that of the SLM process ($2.9 \times 10^{-3 \mu m}$). Consequently, due to the reduced porosity in the EBM-produced part, it is expected to exhibit a higher fatigue limit compared to the SLM-produced part [18].

8.5.2 Applications

Currently, powder bed melting AM technology finds extensive applications in diverse industries. The study [9] provides a compilation of these applications based on the type of metal employed in the process. The medical, aerospace, automotive, electrical, instrument, dental-medical, jewelry, oil and gas industries, among others, all benefit from the utilization of this technology.

SLM technology has gained significant recognition, with one notable application being the construction of the new fuel nozzle design for GE's "GE LEAP" brand. The study [11] provides a visual representation of this component. The development of this nozzle began in 2015, and by 2018, a remarkable 30,000 units had been produced. The material utilized for this piece is a cobalt-chromium alloy specifically designed for aircraft engines.

Furthermore, the GE9X engine model incorporates 28 instances of this innovative nozzle. The engine itself is employed in steam turbines, featuring 228 low-pressure blades. Through the utilization of SLM technology, the number of components required for this nozzle was significantly reduced from 18 to just one, leading to a notable weight reduction of 25% [11]. This accomplishment showcases the immense potential and advantages offered by AM in the aerospace industry.

The study [2] showcases the first successful implementation of a titanium connection for the Airbus A350 XWD airplane using the SLM method.

This connection, being hollow and manufactural with the SLM technique, resulted in a remarkable 30% reduction in weight compared to the previous design. This weight reduction has contributed to lower fuel consumption, potentially increasing the aircraft's cargo capacity. Additionally, the SLM process eliminates the need for tools, leading to a significant cost-saving and a remarkable 75% reduction in production time compared to traditional manufacturing methods, which typically take 6 months or longer [2].

In study [19], orthopedic medical parts such as knee and thigh joints produced using SLM and DMLS methods are displayed. Numbered 1, 2, 3, and 4 represent screws, plates and intramedullary plates, intramedullary rods, knee joints, and thigh joints, respectively. These parts are predominantly fabricated using the Ti6Al4V alloy, which possesses properties that align well with the growth and integration of bones within the human body [19]. This application highlights the immense potential of AM in the medical field, enabling the production of customized implants with biocompatible materials.

8.5.3 Binder Jet

The BJ process, where a model is formed on a powder bed. In this method, a nozzle injects binder material onto the powder bed, following the two-dimensional cross-section of the model's geometry. This process results in the melting and formation of a new layer [4].

The term "Binder" refers to a substance that binds the powder particles together, similar to glue. Typically, binders are in liquid or powder form and possess adhesive properties. The jet binder process can utilize metals (such as stainless steel), polymers, and ceramics as materials [3]. During the process, the powder is leveled using a roller. Once the process is completed, excess powder is removed from the bed, leaving behind a clean model. The model is then subjected to various coatings. The minimum layer thickness in the jet binder process ranges from 0.013 to 0.076 mm. Ceramics, metals, and polymers are among the materials used in AM through the jet binder method. One of the advantages of this process is its ability to produce colored parts. Additionally, the jet binder method eliminates the need for support structures. However, this process has limitations, such as medium to high equipment costs [4].

8.5.4 Direct Energy Deposition (DED)

The Directed Energy Deposition (DED) process offers advantages over other methods due to its freedom in part geometry and absence of dimensional restrictions, making it suitable for producing very large parts. Unlike powder bed melting methods, the DED process does not rely on a powder bed. However, its resolution and accuracy are relatively weaker, with a minimum reported value of around 1 mm. This limitation makes DED particularly well-suited for large-scale part production [11].

According to the ISO/ASTM 52900 standard, the DED process is defined as "An AM process in which concentrated thermal energy is used to melt metals for the purpose of deposition" [20]. The process involves direct energy injection, also known as direct deposition, where a nozzle is positioned on a multi-axis arm to directly inject the material at the desired location. Metals like titanium and chrome-cobalt are commonly used in DED, while polymers and ceramics are not suitable. The nozzle control in this process allows for grain structure control at high temperatures [3]. While laser coating, a type of DED process, was extensively discussed in the previous chapter, this section briefly touches on additional aspects. It's worth noting that lasers, electron beams, and electric arcs can all serve as energy sources for DED, with lasers being the most commonly discussed. Other DED systems, such as arc wire DED and electron beam wire DED, employ wire as an additive while using electric arc and electron beam as the energy sources, respectively. Each process has a different deposition rate, which can influence heating and cooling rates, leading to variations in the metallurgical characteristics, mechanical properties, and thermophysical properties of the part [11]. Notably, the cost of producing parts using the arc wire DED method is generally lower compared to laser DED [21]. To provide a better understanding, the study [20] presents the four conventional categories of DED (Laser Additive Manufacturing-DED (LAM-DED), Wire Arc Additive Manufacturing (WAAM), Wire Laser Additive Manufacturing (WLAM), and Wire and Electron Beam Additive Manufacturing (WEAM)) based on conventional layer thickness, minimum layer width, thermal density, energy efficiency, environmental conditions, layering rate, and layering efficiency, along with additive materials [20].

The study [20] provides valuable insights into the different categories of Directed Energy Deposition (DED) processes. Based on the table, it is evident that the WEAM process exhibits the highest thermal density, resulting in the highest layer thickness, layering rate, and minimum layer width. On the other hand, the WAAM process boasts the highest energy efficiency, reaching 90%. This signifies that a significant portion of the input energy in the WAAM process is effectively utilized for layering and part production.

Recognizing the significance of the WAAM process, the study [20] outlines various applications and corresponding details associated with this method. The table includes information such as dimensions, weight, material type, layering rate, and cost reduction rate. These details provide valuable insights into the implementation of the WAAM process in different scenarios, showcasing its potential for achieving dimensional, weight, and cost-related objectives.

According to the study [20], it can be seen that the WAAM process is carried out in large-scale dimensions with a layering rate of more than 0.8 kg/hour in different materials. NASA is one of the leaders in using the DED process in the manufacture of large-scale parts such as rocket engine components and nozzles. Although the design of nozzles made by the DED method was similar to conventional methods, thanks to AM technology, soldering and assembly problems were eliminated from the manufacturing process. Also, the number of parts was reduced from 1,100 to 10. It is worth mentioning that these parts are usually made using forging and casting with machining operations, and with AM technology, their production has been significantly reduced in terms of cost and construction time.

In 2021, the MX3D company in Amsterdam unveiled the first pedestrian bridge constructed using the Wire Arc Additive Manufacturing (WAAM) process. This remarkable bridge spans a length of 12 m and marks a significant milestone in the implementation of WAAM technology. The construction of the bridge utilized 6 t of stainless steel and involved the use of 4 robots. While the production phase took only 6 months to complete, the design for the bridge was initially conceived by Joris Larman Lab in 2018. In 2019, the final design underwent rigorous testing, subjecting the bridge to a load of 20 tons over a period of 1 month. After successfully passing all relevant tests, the bridge was officially opened to the public in 2021, offering a practical and functional structure.

The study [9] provides visual documentation of the different construction stages involved in bringing this innovative bridge to life. This project showcases the capabilities of WAAM technology in creating large-scale structures and demonstrates its potential for transforming traditional construction processes.

8.5.5 Material Extrude

AM, also known as 3D printing, has revolutionized the manufacturing industry by enabling the production of complex geometries and customized parts with high precision and quality. Among the various techniques employed in AM, one of the most widely used methods is fused deposition, specifically the fused deposition modeling (FDM) process. In FDM, the process involves the rotation of a wire filament around a coil, which is then deposited in the XY direction through a thermal nozzle. The nozzle gradually moves in the Z direction, corresponding to the height of each layer. The layer thickness typically ranges from 0.25 to 0.13 mm, allowing for fine control over the dimensional accuracy and surface finish of the printed part.

The FDM process exhibits remarkable versatility, capable of producing parts with maximum dimensions of up to 914×610×914 mm. This attribute makes it suitable for fabricating a broad range of objects, from small intricate components to large-scale prototypes or even functional end-use parts.

Polymer-based materials dominate the FDM process, with various trade names associated with their specific properties. Commonly used materials include acrylonitrile butadiene styrene (ABS), ABSi, ABS-M30, ABS-ESD7, polycarbonate-ABS (PC-ABS), polycarbonate-ISO (PC-ISO), and ULTEM. Each material possesses distinct characteristics such as strength, durability, heat resistance, or electrostatic discharge protection, enabling designers and engineers to choose the appropriate material for their specific application.

Regarding cost-effectiveness, FDM equipment is moderately priced, making it a viable option for a wide range of users. The process offers the advantage of producing parts with exceptional precision and quality, meeting the requirements of various industries and applications. This level of precision is showcased in the part example presented in studies [3,4], demonstrating the capability of FDM to produce intricate and accurate geometries. In conclusion, FDM is a prominent AM technique widely employed in various industries. Its ability to produce parts with precision, its versatility in terms of size, and the availability of a wide range of polymer-based materials contribute to its popularity. As technology advances and new materials are introduced, the FDM process continues to play a pivotal role in AM, enabling the creation of complex and functional parts with excellent quality and reliability.

8.5.6 Vat Polymerization (VP)

The vat polymerization process is a technique employed in AM, specifically for creating solid objects using photopolymer resin in a liquid state. In this process, a tank containing the liquid resin is utilized. A frozen layer of ultraviolet (UV) light is gradually projected onto the surface of the liquid resin, causing it to solidify and form a three-dimensional object. Each layer in the vat polymerization process typically has a thickness ranging from 0.1 to 0.2 mm. Acrylic resins, such as Poly1500, are commonly used materials in this process due to their favorable characteristics for photopolymerization. The equipment and materials required for vat polymerization are considered to be of medium to high cost. This is due to the specialized nature of the machinery and the need for UV light sources and precise control mechanisms. However, the expense is justified by the high level of accuracy achievable in this process.

Study [3] serves as an example to demonstrate the exceptional precision and quality that can be achieved through vat polymerization. The process is capable of producing intricate and detailed structures with fine features and smooth surface finishes. It is important to note that producing internal holes or voids in the vat polymerization process can present challenges. The solidification process may result in the destruction or deformation of base areas or supports within the object. Therefore, careful consideration and additional support structures are often required when designing objects with internal cavities.

In summary, the vat polymerization process is an AM technique that utilizes liquid photopolymer resin to produce solid objects. It offers high accuracy and the ability to create complex geometries with fine details. Although it requires specialized equipment and materials, the process is capable of delivering exceptional results. However, the production of internal holes can be challenging due to potential damage to the base areas. Nonetheless, vat polymerization continues to be a valuable technique in AM, contributing to advancements in various industries.

8.5.7 Material Jet

The material jet process is an AM technique that combines polymers, plastics, and sodium hydroxide solutions to fabricate parts. In this process, the materials are jetted onto the cross-sectional surface in the form of droplets through a specialized nozzle. These droplets are then solidified using two UV rays along their trajectory. The minimum thickness of each layer in the material jet process is reported to be 0.017 mm, allowing for the creation of fine and detailed features. An example of a part produced using this method is shown in a study [3], showcasing the quality and suitability of the final product. The material jet process is known for its high production accuracy, enabling the fabrication of precise components. However, it is important to note that the materials and equipment required for this process are relatively expensive. The specialized nature of the machinery, coupled with the use of specific materials and UV curing technology, contributes to the cost factor.

One challenge associated with the material jet process is the removal of the support structures used during fabrication. These supports are necessary to provide stability and prevent deformation during the printing process. However, their removal can be problematic, particularly when attempting to produce hollow parts. The presence of support structures within the hollow areas can make it difficult to clean and achieve a fully functional hollow structure.

In summary, the material jet process is an AM method that utilizes polymers, plastics, and sodium hydroxide solutions to create parts. It involves jetting material droplets onto the cross-sectional surface and solidifying them using UV rays. The process offers high accuracy and produces parts with suitable quality. However, the materials and equipment involved are relatively expensive. Additionally, producing hollow parts can be challenging due to the difficulties associated with removing support structures. Nonetheless, the material jet process continues to be an important technique in AM, contributing to advancements in various industries.

8.5.8 Sheet Lamination

The process of SL in AM encompasses two distinct types: ultrasonic additive manufacturing and AM of sheet objects (LOM). In the LOM, employing a combination of laser, plate, and glue to generate laminated sheets. Through laser-guided glue injection, the desired model gradually takes shape at specific locations on the page. Each layer produced in this process measures 0.165 mm in thickness, while the maximum dimensions of the resulting model are reported to be 170×220×145 mm.

The LOM process offers an advantage in terms of affordability, with the equipment required being moderately priced, and the raw materials used generally being inexpensive.

8.5.9 Comparisons of Methods

The previous sections have presented various AM processes, which can be compared from different perspectives. The study [9] provides a comparison of the powder and wire-based Directed Energy Deposition (DED), PBF, and BJ processes in terms of production part dimensions and accuracy. Based on the figure, it is evident that the wire and powder DED processes are suitable for manufacturing large-scale parts where precision is not a critical factor. Conversely, the BJ and PBF processes are preferred for producing highly accurate and often smaller parts.

To gain a better understanding, the study [14] compares the laser-based DED and PBF processes from various viewpoints. As per the data presented in the study [14], it can be observed that the layering speed of the PBF laser process is slower compared to DED. However, the surface quality and accuracy of parts produced using the laser PBF method are significantly higher than those produced using DED. Consequently, the DED process is commonly employed in the production of large-scale parts. To summarize and compare all seven AM processes presented in this chapter, a study [22] is provided. Overall, these comparisons provide insights into the strengths and limitations of each AM process, allowing for informed decisions based on specific production requirements.

REFERENCES

[1] S. C. Altıparmak and B. Xiao, "A market assessment of additive manufacturing potential for the aerospace industry," *Journal of Manufacturing Processes,* vol. 68, pp. 728–738, 2021.

[2] R. Liu, Z. Wang, T. Sparks, F. Liou, and J. Newkirk, "Aerospace applications of laser additive manufacturing," In: *Laser Additive Manufacturing.* Elsevier, pp. 351–371, 2017.

[3] D. D. Singh, T. Mahender, and A. R. Reddy, "Powder bed fusion process: A brief review," *Materials Today: Proceedings,* vol. 46, pp. 350–355, 2021.

[4] M. Jiménez Calzado, L. Romero, I. A. Dominguez Espinosa, M. d. M. Espinosa, and M. Domínguez Somonte, "Additive manufacturing technologies: an overview about 3d printing methods and future prospects," Complexity in Manufacturing Processes and Systems 2019, Complexity, Hindawi, pp. 1–30, 2019.

[5] G. Liu *et al.*, "Additive manufacturing of structural materials," *Materials Science and Engineering: R: Reports,* vol. 145, p. 100596, 2021.

[6] Z. Zhu, V. G. Dhokia, A. Nassehi, and S. T. Newman, "A review of hybrid manufacturing processes-state of the art and future perspectives," *International Journal of Computer Integrated Manufacturing,* vol. 26, no. 7, pp. 596–615, 2013.

[7] B. Redwood, F. Schffer, and B. Garret, *The 3D Printing Handbook: Technologies, Design and Applications.* 3D Hubs, 2017.

[8] O. Abdulhameed, A. Al-Ahmari, W. Ameen, and S. H. Mian, "Additive manufacturing: Challenges, trends, and applications," *Advances in Mechanical Engineering,* vol. 11, no. 2, p. 1687814018822880, 2019.

[9] J. Hart, *An Introduction to Metal Additive Manufacturing.* Massachusetts Institute of Technology: MIT Lecture Note, Massachusetts Institute of Technology MIT Lecture Note, October 18, 2021.

[10] B. DeBoer, N. Nguyen, F. Diba, and A. Hosseini, "Additive, subtractive, and formative manufacturing of metal components: a life cycle assessment comparison," *The International Journal of Advanced Manufacturing Technology,* vol. 115, no. 1–2, pp. 413–432, 2021.

[11] B. Blakey-Milner *et al.*, "Metal additive manufacturing in aerospace: A review," *Materials & Design,* vol. 209, p. 110008, 2021.

[12] R. Housholder, "Molding Process, 1981," Patent# US4247508A.
[13] W. E. King *et al.*, "Laser powder bed fusion additive manufacturing of metals; physics, computational, and materials challenges," *Applied Physics Reviews,* vol. 2, no. 4, 2015.
[14] T. Duda and L. V. Raghavan, "3D metal printing technology: the need to re-invent design practice," *Ai & Society,* vol. 33, no. 2, pp. 241–252, 2018.
[15] L. S. Bertol, W. K. Júnior, F. P. Da Silva, and C. Aumund-Kopp, "Medical design: Direct metal laser sintering of Ti-6Al-4V," *Materials & Design,* vol. 31, no. 8, pp. 3982–3988, 2010.
[16] V. Bhavar, P. Kattire, V. Patil, S. Khot, K. Gujar, and R. Singh, "A review on powder bed fusion technology of metal additive manufacturing," 4th International conference and exhibition on Additive Manufacturing Technologies-AM-2014, September 1 & 2 ,2014, Banglore, India, pp. 251–253, 2017.
[17] L. Löber *et al.*, "Comparison off selective laser and electron beam melted titanium aluminides," In: *2011 International Solid Freeform Fabrication Symposium.* University of Texas at Austin, 2011.**
[18] X. Zhao *et al.*, "Comparison of the microstructures and mechanical properties of Ti-6Al-4V fabricated by selective laser melting and electron beam melting," *Materials & Design,* vol. 95, pp. 21–31, 2016.
[19] F. Fina, S. Gaisford, and A. W. Basit, "Powder bed fusion: The working process, current applications and opportunities," *3D Printing of Pharmaceuticals,* 81–105, 2018.
[20] D.-G. Ahn, "Directed energy deposition (DED) process: State of the art," *International Journal of Precision Engineering and Manufacturing-Green Technology,* vol. 8, pp. 703–742, 2021.
[21] F. Wang, S. Williams, P. Colegrove, and A. A. Antonysamy, "Microstructure and mechanical properties of wire and arc additive manufactured Ti-6Al-4V," *Metallurgical and Materials Transactions A,* vol. 44, pp. 968–977, 2013.
[22] C. Sun, Y. Wang, M. D. McMurtrey, N. D. Jerred, F. Liou, and J. Li, "Additive manufacturing for energy: A review," *Applied Energy,* vol. 282, p. 116041, 2021.

Index

Note: **Bold** page numbers refer to tables and *italic* page numbers refer to figures.

Printed in the United States
by Baker & Taylor Publisher Services